In this unique book, Sir Donald Harrison draws on his wide-ranging experience as a surgeon and comparative anatomist to produce an authoritative and detailed account of the anatomy and physiology of the mammalian larynx. His investigation of the larynx has involved the study of over 1400 specimens of mammalian larynges from around the world, as well as using data from his own clinical experiences. The comparative morphology of the larynx is discussed from a developmental and functional perspective, and the involvement of the larynx in respiration, locomotion and vocalization is highlighted. Throughout the book the relationship of structure to function is drawn out, and the clinical relevance of features of the human larynx is emphasized.

This book will be an invaluable reference for all researchers and clinicians involved in laryngology, as well as for anatomists, zoologists and anaesthesiologists.

The Anatomy and Physiology
of the Mammalian Larynx

The
Anatomy and Physiology
of the Mammalian Larynx

D F N HARRISON

University of London

CAMBRIDGE
UNIVERSITY PRESS

CAMBRIDGE UNIVERSITY PRESS
Cambridge, New York, Melbourne, Madrid, Cape Town, Singapore, São Paulo

Cambridge University Press
The Edinburgh Building, Cambridge CB2 2RU, UK

Published in the United States of America by Cambridge University Press, New York

www.cambridge.org
Information on this title: www.cambridge.org/9780521453219

First published 1995

A catalogue record for this publication is available from the British Library

Library of Congress Cataloguing in Publication data

Harrison, D. F. N. (Donald Frederick Norris)
    The anatomy and physiology of the mammalion larynx/D. F. N. Harrison.
      p. cm.
    Includes index.
    ISBN 0-521-45321-6 (hardback)
    1. Larynx–Anatomy. 2. Larynx–Physiology. 3. Anatomy,
Comparative. I. Title.
    [DNLM: 1. Larynx–anatomy & histology. 2. Larynx–physiology.
    3. Mammals. 4. Anatomy, Comparative. 5. Physiology, Comparative.
WV 501 H318a 1995]
QM331.H37   1995
599′.012–dc20
DNLM/DLC
for Library of Congress   95–742   CIP

ISBN-13 978-0-521-45321-9 hardback
ISBN-10 0-521-45321-6 hardback

Transferred to digital printing 2006

Dedicated to my wife Audrey,

*who has understood and accepted my determination
to combine a surgical career with that of an amateur comparative
anatomist. Without her support and tolerance this book would
never have been completed.*

Contents

Preface

In 1929 Victor Negus, a Junior Surgeon for Diseases of the Throat and Nose, Kings College Hospital, London, published a book entitled *The Mechanism of the Larynx*. Part of this was based on his research for a Degree of Master of Surgery in the University of London. However, this 528 page volume contained unique observations on the comparative morphology of a wide range of animal larynges, cumulating in a fundamental account of the origin, development and mechanism of this unique organ.

Early in my own laryngological career, I had read and marvelled at the erudition of this 'amateur comparative anatomist', ultimately hoping to extend his research by utilizing modern techniques and a wider range of animals. As Professor of Laryngology and Otology at the Institute of Laryngology and Otology in London, I had developed a laboratory for sectioning human larynges removed for cancer. This was for an investigation into the manner by which neoplasia spreads through this organ. On completion of this research in 1976, these facilities became available for detailed examination of other mammalian larynges. Specimens were collected over the following 17 years from zoological institutions, game parks and helpful friends throughout the world. Most, however, came from post-mortems carried out in the Pathology Department of the Zoological Society of London. Without the enormous support and good will of this institution and many other helpful people, this unique collection of mammalian larynges would have been impossible; my gratitude is infinite.

With over 1400 specimens and 100 cot death larynges, the preparation and processing of all this material required the skilled and enthusiastic support of a small number of dedicated individuals. Most of the sectioning was carried out by Mrs Christine Brock, whose technical expertise with rare and unusual materials is unsurpassed. Several medical students working for their BSc in Anatomy, or just interested in comparative anatomy joined in, Jonathan Blaxhill and Anna Crown (now Dr Staricoff) being especially notable. Foremost, however, was my elder daughter Susan Denny. A zoological graduate, she has been responsible for the major part of the data collection, illustrations, museum preparations and numerous other tasks, which my occupation with the problems of head and neck cancer prevented my undertaking. At night I assembled all this material, examined slides, made notes and then transferred the data to custom designed punch cards. Eventually, having taught myself some skill in the use of spread sheets, data bases and statistical analysis, everything was stored on a hard disc, awaiting the day when I could devote enough time to digest this information. Retirement meant cessation of surgery but not a great deal of spare time; although now I could begin to collect my thoughts and generate a plan for the compilation of the book I had wanted for several years to write.

Cambridge University Press agreed to undertake the printing of a book whose subject might well have been considered as 'unfashionable'; for comparative anatomy even when related to a clinically important organ, is not as popular today as in the past. For their support and encouragement I remain eternally grateful. My successor at the Institute, Tony Wright, generously provided space for the extensive collection of slides, museum pots and fixed specimens. The Department of Clinical Photography at this institution were responsible for all the illustrations and again, I am most appreciative of their unrivalled expertise. A valued colleague and friend Robert Pracy, MPhil., agreed to read the draft manuscript to 'tidy up' the grammar and punctuation. Any residual errors, however, are solely due to my ignoring his wise council!

Despite considerable experience in medical writing, the compilation of a more scientifically orientated book proved surprisingly difficult, although a great deal more interesting. Much new data had to be digested, analysed and then correlated with existing knowledge of the anatomy and physiology of the mammalian larynx. This, however, highlighted considerable gaps existing in our knowledge of the fundamental processes that govern laryngeal function. Morphological studies, whilst providing valuable three-dimensional information, emphasize the variance that occurs within species and between individual orders. Unfortunately, any extension of similar studies, designed to include larger numbers of more unusual or exotic species, remains unlikely because of the scarcity of funds and the difficulty of obtaining specimens in a changing environment. With these limitations this work, although by a self-confessed 'amateur' anatomist, will probably remain as a unique investigation into the structure and function of one of Mammalia's most fascinating organs. Hopefully, it will be of interest to many disciplines, and at the same time stimulate an increased interest in comparative morphology.

D. F. N. H.

1 Introduction

The development of comparative anatomy has been neither progressive nor continuous. Progress has been halted by intellectual stagnation, by failure to capitalize on earlier discoveries and by religious and political intolerance. Rarely can major advances be attributed to individuals but are usually the outcome of centuries of human endeavour. The eight centuries of Greek objectivity were followed by a period of scientific and artistic stasis which lasted for more than a thousand years.

It is unclear as to whether the earliest studies of animal anatomy were based primarily on a desire to study animal structure *per se*, or because of a hope that light might be thrown on the morphology of man. Animal dissections by both zoologists and anatomists date at least from the time of Aristotle (*c.* 384 BC) and reveal an objectivity common with other facets of Greek culture. However, the foundations for much of this scientific thought were developed in the earlier civilizations of Babylon and Egypt and were then developed by Greek philosophers primarily to assist and explain human problems. Aristotle's contributions to both zoology and comparative anatomy were extensive and his *History of Animals, the Parts of Animals and the Generation of Animals* establish him beyond question as the founder of the biological sciences. Descriptions of the nictitating membrane of the eye, os cordis of the horse and ox and the os penis found in many carnivores were based upon his own dissections, although subsequent errors in translations of his works have seriously reduced our knowledge of the actual number of animals Aristotle studied. In common with other Greek scientists of his time, his work was based on observation but he founded the method of biological assessment which in essence is still in use today. First, statement of the problem, then discussion of previously published work before finally drawing conclusions. Even so, factual errors occur such as his belief that the wolf and lion have only one cervical vertebra. Although dividing the animal kingdom into the equivalent of Invertebrata and Vertebrata with recognition that Cetacea were not fishes, he made no direct attempt to draw up a comprehensive classification of species based on specific criteria, although his writings suggest that he was moving towards a *natural* classification. His wide ranging interests certainly provided much anatomical information, although it remains unclear as to how much was original and how much came from studies made by others. Despite his absorption with all aspects of human life, however, there is no record of his dissecting the human body, possibly reflecting the social and religious mores of the times.

Galen (*c.* AD 130–201) lived when Greek culture was in decline. His interests, in contradistinction to Aristotle, lay in the morphological and physiological aspects of biology. As an observer and experimentalist much of his writing is said to represent the anatomical traditions of his time and thus perpetuate many errors of fact, whilst at the same time perpetuating his preeminence in the field of biology for many centuries. He founded a system of physiology based upon experiments carried out on living animals, this included observations that in the living animal the left ventricle and arteries generally contain blood and not air, thus correcting a belief that had existed for over three hundred years. His demonstration that nerves originated in the brain and spinal cord rather than the heart was again a fundamental discovery; but these fundamental discoveries were combined with a mixture of established facts and misconceptions, many of which were to persist well into the twelfth century.

There is reliable evidence that Galen dissected a wide variety of animals including numerous mammals. Amongst these were both tailed and tailless primates and as was common at this time he transferred by analogy his conclusions to 'Man'. As physician to the gladiators in Pergamun, Galen must have been in an excellent position to observe much human anatomy, both internal and external, but recorded few of his conclusions (Cole, 1944).

With the death of Galen in AD 200 and the gradual decline in Greek culture and philosophy, the tradition of biological learning virtually ceased in Europe. Translations of texts into Latin in the eleventh and twelfth centuries resulted in blind acceptance of what was in essence, outdated dogma. For many centuries Mankind was to be either indifferent to the mysteries of Nature or attempt to penetrate them by a mixture of pious belief and literary research. The revival of learning, when it arrived, was due to the invention of movable type rather than any change in philosophical beliefs. Prior to this the diffusion of knowledge had been gradual and restricted to hand copied manuscripts usually compiled within monasteries, where the Abbot might be expected to be watchful for potential heresies! It is hardly surprising therefore that between the eclipse of classical learning and the invention of printing in the fifteenth century, anatomical research was largely in abeyance. Scientific research of all kinds was considered unnecessary, even forbidden, and Greek teachings were accepted as authoritative by University and Church.

Standing virtually alone during this barren period was the majestic figure of Leonardo da Vinci (1452–1519), whose interest in comparative anatomy was the result of his work as painter and sculptor. He combined the detailed objectivity of the anatomist with the vision and aesthetic sense of an artist and was greatly interested in the general structure of the human. When mystified despite having dissected more than 30 human bodies, he turned to comparative anatomy for solutions. Had his findings been published at the time they would surely have had a profound influence on biological thought. However, he was denounced by the Pope for the practice of human anatomy, his contemporaries bitterly opposed many of his discoveries, such as 'that the heart was a muscle', and his manuscripts only became widely available 200 years later. His techniques are still partially known and include the making of wax casts of the cerebral ventricles and the use of serial sections. He dissected many animals including primates and his descriptions of the facial muscles of the horse remain masterpieces of accuracy (Hopstock, 1921).

Prior to the development of movable metal-type, printing manuscript copies of documents was slow and laborious. The biological works of Aristotle were said to have been copied more extensively than those of any other scientist during these Dark Ages, yet few remain in existence. The first of the great biological works to be published during the sixteenth century was the *Fabrica* by Vesalius in 1543. Although possibly the 'Father' of modern anatomy he was not a comparative anatomist, being analytical and unimaginative rather than constructive in outlook. His contribution to progress was in establishing the anatomical method by which he showed that progress in science was possible only by research. Authority, no matter how firmly established, must give way to the findings of original investigation; by this dictum he generated implacable opposition. He accepted the authority of the human body as revealed by dissection, rather than relying on Galen's view which was derived largely from animal dissection. 'Galen', said Andreas Vesalius, 'imposed upon us the anatomy of the ape.' Even so Vesalius's debt to Galen was considerable, for the former's physiology is but a revised version of Galen's earlier experimental system. Apart from the illustrations, the *Fabrica* is based largely on Galen with its text Grecian in outlook. Factual errors that were said to have been imported from animal anatomy were thought to have been corrected, but many new ones were left uncorrected. Notwithstanding this, *Fabrica* remained the most important anatomical reference book of its day and its author the leader of the scientific revolution of the sixteenth century. Embittered by local opposition, Vesalius gave up scientific research to become personal physician to the Spanish monarchs Charles V and Philip II. He was accused of opening the chest of a supposedly dead nobleman only to find the heart beating! For this he was sent to Jerusalem as a penance, was shipwrecked on the return journey and died in 1564 on the Greek island of Zakinthis (Dunning, 1993).

The revival of research in comparative anatomy may be dated from the publication in 1551 by Pierre Belon on various '*Cetacea and other Marine animals*'. Although dissecting three species of cetaceans and noting many of their mammalian characteristics he related them to the fishes. Similarly bats were put amongst the nocturnal birds of prey. It was inevitable that such early studies in comparative anatomy would be deficient in morphological accuracy, for knowledge of descriptive anatomy was itself at a primitive stage forestalling significant comparisons. More important than Belon was Frisian Volcher Coiter whose works were published between 1572 and 1575 (Cole, 1944). A student of the Fallopian School of Padua he urged the comparison of human anatomy with that of beasts, and although well versed in human anatomy his prime interest lay in dissections of a wide variety of mammals, amphibians and aves. His researches into the growth of the skeleton in the human foetus showed that bones are preceded by cartilages and the importance of centres of ossification. Although his studies embraced a wide range of animals his main contribution to mammalian anatomy was based on studies of the skeletons of more than 50 species, including both tailed and tailless primates. The first comprehensive monograph devoted to a single species, the horse, was published in 1598 by Carlo Ruini (a senator of Bologna) one month before his murder. There is suspicion that the plates used in this monograph were those drawn by da Vinci for his own proposed treatise but prior to this publication the anatomy of the horse was largely unknown, despite it being such a well known domestic animal. This work was the logical outcome of the Vesalian tradition, topographical but logical in its schematic approach to the horse's system. It passed through 15 editions between 1598 and 1769 despite extensive plagiarism and prejudice. Dissection of such a large unpreserved animal must have presented many technical problems but was made possible by Ruini's knowledge of human anatomy. However, the absence of a comparative bias hindered the identification and naming of the musculature and he failed to recognize the absence of a clavicle.

Records of the anatomy school at Padua go back at least to 1387, although its most brilliant period was in the sixteenth and seventeenth centuries. During this time the famous Chair in Anatomy was occupied by Vesalius, Columbus, Fallopius and Fabricius. The latter, a convinced and ardent Aristotelian and

Galenist, attempted to combine scholasticism with research, although both approaches to furtherance of knowledge were mutually exclusive. Although primarily interested in comparative embryology he must be credited for being the first to investigate the physical importance of animal structure. In 1594 he built, at his own expense, the famous 'Anatomy Theatre' to accommodate his large classes and it was here that he first dissected publicly the human body.

Many of Fabricius's publications on embryology and morphological anatomy compare the human with other animals, assessing differences and factors in common both in structure and function (e.g. Fabricius, 1600). Studies included the alimentary canal and later the compound stomach and physiology of rumination in animals such as the sheep, goat, ox and deer. Other comparative studies included the eye, respiratory system and ear. If the work of Fabricius lacked philosophical interest it undoubtedly exercised a considerable influence on the development of anatomical science. Vesalius had concentrated on human anatomy to the exclusion of comparative mammalian anatomy. Fabricius by stressing the importance of comparative studies restored interest in animal dissection influencing his successor Casserius who in 1600 promulgated his classic investigations on the organs of sense and voice (Casserius, 1601).

The anatomy of the vocal and respiratory tract

Preeminent amongst the earlier anatomists in his interest in the physiology of the nervous system was Galen. Although only recognizing seven of the cranial nerves he compared the recurrent laryngeal nerves in several species including ox, dog and bear. Sensory and motor nerves were distinguished by variations in texture but his investigations into the neurological control of muscles led to research into the production of voice by the larynx and the function of respiratory muscles. This interest in the larynx was continued by Leonardo da Vinci who examined various carnivore larynges recording his findings with commendable accuracy. Vesalius demonstrated that cutting the recurrent nerves 'silenced the voice' but was unable to dissect the brain because of ecclesiastical pressure. His experiments on the respiratory movements of the chest, lungs and diaphragm were performed on dogs and pigs, virtually his only interest in comparative anatomy. He had criticized Galen for introducing animal anatomy into his illustrations of human anatomy but the figure of the hyoid bone in his own treatise comes from the dog and the recurrent laryngeal nerve is not human.

Belon (1551) discovered the function of the intranarial epiglottis in cetaceans, understanding that despite an aquatic life they were transnasal air breathers. However, the most detailed account of mammalian laryngeal anatomy is by Ruini (1598) in his classic description of the horse. Cartilages, muscles and epiglottic function are all accurately recorded as is the course of the recurrent laryngeal nerves, although the latter are erroneously thought to be branches of the spinal nerves. In 1600 Fabricius wrote full accounts of the comparative anatomy of the laryngeal region of several mammals including the ape but tried to homologize the laryngeal cartilages of birds and mammals (Figure 1.1). Later he entered the difficult field of phonetics discussing the physiology of voice production, functions of laryngeal structures and the possible reasons for the anatomical placement of the larynx. The speech of humans is compared with animal vocalization using air blown through an excised larynx to illustrate the importance of the vocal cords in sound production. Considering his poor reputation as a comparative anatomist these investigations were of considerable importance, providing considerable stimulus to his successor Guiulio Casserius. Despite his humble background (Casserius was man-servant to Fabricius) his publications on the sense and voice organs (Casserius, 1601) represented the most ambitious and detailed investigations on comparative anatomy carried out at this time. Although limited in scope, his dissections illustrated the anatomy of the larynx and chest in the human, ape and more than 20 other mammals. The recurrent laryngeal nerves were now correctly assigned to the sixth pair of cranial nerves and their pathway around the subclavian on the right and aorta on the left correctly described. The larynx was recognized as the principal organ of voice production, the cartilages and intrinsic muscles clearly identified. A long philosophical discussion on the definition, nature and cause of voice was unaccompanied by any experimental evidence although the larynx is identified as the source of sound rather than the movement of the lungs. Extension of this interest in sound production to the invertebrates provided the first account of the timbale found on the abdomen of the cicada and he was aware that in some crickets stridulation is produced by friction of areas on wing surfaces.

William Harvey (1578–1657)

Early anatomists were preoccupied with human anatomy despite the difficulties and perils attached to human dissection. Cadavers were difficult to obtain and the study of human anatomy was viewed by an unenlightened public with disgust and by the Church as anti-religion. In the seventeenth century the emphasis on experimentation and measurement with increasing conviction that the body

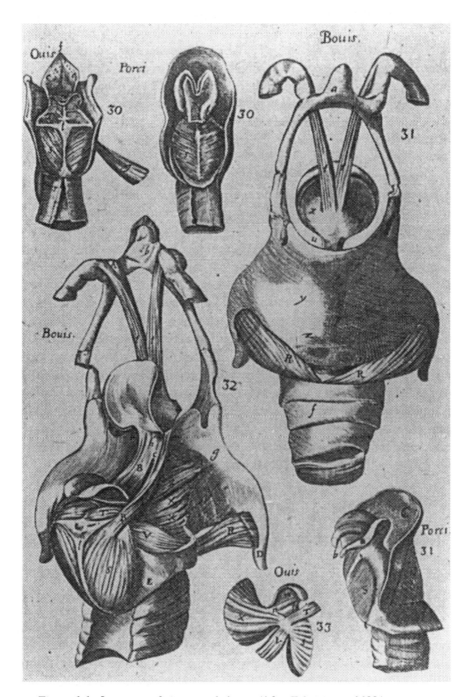

Figure 1.1. Larynges of pig, ox and sheep. (After Fabricius, *c.* 1600.)

could be understood by mechanical principles, were expressed through a preoccupation with bodily mechanisms. The embryological explanation for the variations in routing of the recurrent laryngeal nerves was still unknown and William Harvey developed a mechanistic theory of voice production based on supposed similarities with cords stretched around pulleys. Although visualizing the need for research, he remained misled by the tyranny of the Greek tradition and engaged in a long philosophical exposition on the nature and causes of voice production without carrying out any experiments to support his suppositions. Nevertheless the prophetic genius of Harvey saw the real value of research in non-human mammals in that it overcame the inherent difficulties of investigating the relationships between structure and function in humans. He applied this method most successfully to his studies on the function of the mammalian heart (Harvey, 1628).

Anatomy began to progress again, just as it had following publication of the *Fabrica* of Vesalius and by the end of the seventeenth century a considerable body of reliable information within the biological sciences had been established. Marco Severino, a close contemporary of Harvey, published his treatise *Zotomia Democritaea* in 1645. When comparing the anatomy of the human and the ape he was so impressed by the obvious affinity that he recommended that the ape should be used for all medical research, although as an anatomist he preferred the pig! His advocacy for repeated and varied animal dissections was based on the premise that 'structures might be well developed in one species but absent or less developed in another'. However, in common with most other comparative anatomist-surgeons, identification of the more unusual species that he described remains in doubt and his publications consist largely of brief accounts of often unrelated and sometimes inaccurate anatomical facts.

The new age of comparative anatomy

Acceptance that progress could only be made by combining research with accurately observed morphological information, led to the accumulation of considerable amounts of new and reliable data. In time this formed the basis for a system of comparative anatomy whose foundation had been laid by Ruini at the end of the sixteenth century (Cole, 1944). The most distinguished members of this monographic school of animal anatomy now included Malpighi, Tyson and Swammerdan who in their concern for anatomical investigation rather than philosophical discourse filled the transition period between William Harvey (1578–1657) and John Hunter (1728–93). Malpighi was born in 1628, the year which saw the publication of Harvey's work on the circulation. In 1661 Malpighi demonstrated the true nature of the lungs and with the discovery of the blood capillaries, laid the foundation of our knowledge of the physiology of respiration. His most important publications are thought to have been devoted to the structure and life history of *Bombyx mori*, a silk-moth. Tyson however, was primarily interested in the respiratory system of the Cetacea describing the special adaptions of their larynx necessary for an aquatic but air-breathing existence (Tyson, 1680). He concluded 'that it was somewhat inserted into the bottom of the nasal passage and that it was different from other animals', and observed that the blow-hole corresponded to the nostrils of other animals. This was based on studies of the porpoise and supplemented the findings of others, for the Cetacea appear to have attracted considerable interest for more than a century (Belon, 1551).

Although Tyson studied a wide variety of unusual mammals, such as the peccary, noting the dorsal scent gland, its compound stomach and absent gall bladder, he also gave the first account of the anatomy of a marsupial, the opossum. Besides these valuable contributions to comparative anatomy his most important publication was a monograph on a young 'Pigmie', which he proved was not an undeveloped human or monkey (Tyson, 1699). At this time the anatomy of the higher anthropoids was unknown, even the appearance of these animals was familiar only to a few explorers whose accounts attracted general disbelief. The gorilla was first discovered in 1847 and although the orang-utan had been described in 1658 its anatomy was not detailed until 1778, when Pieter Camper identified it as Tyson's 'Pigmie' (Camper, 1778). The chimpanzee 'emerged' in 1625 and was examined, though not dissected, by Tulip who, despite Rembrandt's celebrated 'Anatomy lesson of Dr Nicolaas Tulip' painted in 1632, was a local politician and successful physician rather than a comparative anatomist (Dunning, 1993).

Tyson must be credited for initiating studies into these human-like apes thus recognizing the existence of a new species of primate intermediate between humans and monkeys. His problem was to determine the status of these higher anthropoids and following a decision that the ape was indeed a true species, compiled a list of their most significant anatomical features. From this it was possible to assemble groups of characters which served to emphasize an affinity between different species, a system not entirely dissimilar from that used today by zoologists. Despite his dominant role in seventeenth-century medicine and the volume of animals dissected during his research on the brain, Thomas Willis made little contribution to our knowledge of mammalian comparative anatomy. What little was recorded in his *Soul of Brutes* (1672) appeared to be surplus to his arguments which themselves bore a close relationship to the more pedantic forms of classical Hellenist philosophy. However, this was the period during which Dutch science was paramount throughout Europe and it was their scientists who were to make great contributions to the progress of comparative anatomy. Foremost was the amateur Antony van Leeuwenhoek (1632–1723), who despite an unrestrained desire to place his preliminary and often un-digested thoughts in print, must be credited as the originator of microdissection as an essential tool of anatomical research. A notable feature of his work was that having made an important observation in one animal he would then search for a similar feature in other species, despite his belief in the fixity of species. His manipulative skill must have been astonishing, for although it is recorded that he used more than 400 microscopes which magnified up to 200 diameters, the lens was single and biconvex. Although not considered an originator of many comparative anatomical discoveries, he was particularly interested in aphids and gnats, his microscopical approach was revolutionary in his day setting the scene for those who would follow. Jan Swammerdam, the son of a prosperous apothecary in Amsterdam, was in contradistinction trained in science. Born five years after Leeuwenhoek, he spent much of his early years unprofitably cataloguing the family museum before commencing his most important work on respiration which was published as his Doctorate dissertation in 1667. This became one of the classics in the history of physiology, describing experimen-tally the mechanism of mammalian respiration. However, most of his publica-tions relate to Invertebrata, again using the microscope. Sadly he drifted towards the unrewarding distractions of mysticism and spiritual exaltation, dying it is said 'in the turmoils of an unbalanced mind convinced that the pursuit of natural knowledge was vain and impious'. Yet his work on respiration and the treatises on insects remain a monument to his genius.

Learned societies and institutions

The addition of research to the simple collection of scientific facts eventually resulted in a degree of co-operation between individuals with similar interests and of like mind. In London this may well have been the intention of Boyle's 'Invisible College', which in 1660 became the Royal Society. The main objective of this learned society was the 'advancement of experimental learning by organized research', although at the same time members were reminded to remain aware of the decorum expected of their scientific profession (Birch, 1766). From its foundation, members showed great interest in anatomical studies although it appears from the *Royal Society Transactions* this interest was primarily medical rather than purely biological. Early volumes included descriptions of the morphological anatomy of 22 mammals and these included the cetaceans, deer and several Carnivores. However, no obvious attempt appears to have been made to compare or contrast these anatomical findings between species or with humans. This was in sharp contrast with the French Academy of Science, founded in 1666, where a School of Morphology was established especially for the purpose of studying comparative anatomy. This Academy divided its members into mathematicians who met on Wednesdays, and physicists (which included biologists) who met on Saturdays. The latter's meetings consisted of dissections followed by discussion which in the case of one of their most prominent figures, Claude Perrault, led to his death from infection caught when dissecting a camel! Many of the specimens came from the Royal Menagerie in Paris and in 1732 the Academy published a three volume account of dissections of 49 species of fish, amphibia and birds together with 25 mammals (Gouye, 1688; Perrault, 1734).

In other European countries anatomical studies were blossoming and many founded their own scientific societies. The introduction of foreign membership or correspondents inevitably led to increased exchange of ideas and information. Despite this new awareness, scientists still had to rid their minds of the perverse arguments from opponents who resisted any new idea not based upon established dogma. The Church remained a force not to be ignored, as was to emerge in the following century with the publication of Charles Darwin's *On the Origin of Species*.

The anatomical museum

Present-day acceptance of the essential role played by museums in the development of anatomical studies is a far cry from the problems faced by medieval museums. Lack of efficient means of fluid preservation meant that specimens had to be dried or restricted to objects of a non-perishable nature such as skeletons, horns, fossils, etc. Injection and corrosion techniques made it possible to collect some anatomical material but required considerable technical skill. Even after the introduction of alcohol as a means of obtaining a permanent moist preparation, its use was restricted by the high cost of both spirit and glass. Even the Hunterian Museum had only one-third of its 13,000 specimens in a fluid preservative (Cole, 1944). There is some evidence to suggest that brandy was used as a preservative by Swammerdam (*c.* 1670) but other substances such as salt and alum were tried before it became generally accepted that only spirit of wine possessed adequate anti-putrefactive powers for long-term storage.

However, alcohol has always attracted the attention of taxation and a duty on spirits was first imposed in England in 1643 being rated at six pennies per gallon. If this was not enough to hinder museum development, the invention of flint glass in the seventeenth century producing the colour and transparency needed for display jars, also attracted duty. With greater use this rose from 9s 4d per cwt in 1745 to £1 12s 8d in 1803. It was at the beginning of the eighteenth century that both spirit and flint glass began to be generally used in anatomical museums and it may be that John Hunter used the facilities of the 'spirit-runners' as often as he did the Resurrectionists!

Zoological Society of London and Sir Richard Owen

Interest in animals was not, however, confined to zoologists and anatomists, for exploration had drawn attention to the wide variety of species to be found outside of Europe. Stamford Raffles, following his retirement as Governor of Java in 1824 contacted relatives in London with the express purpose of establishing a Grand Zoological Collection together with a scientific society for the study of living animals. The Linnean Society established in 1788 under the guidance of Joseph Banks was primarily concerned with Botany and there was both interest in and a need for an organization devoted primarily to zoological matters. In 1827, one year after Raffles had died, the Zoological Society of London was formed with a lease of land from the newly established Regent's Park, the gardens being opened to the fellows on the 28th November 1827 (Huxley, 1981).

The Charter of the Society granted by George IV named five Founding Members with the purpose of the society being 'the advancement of zoology and animal physiology and the introduction of new and curious subjects of the Animal Kingdom' (The Charter and Bye-Laws, 1958). Exotic species were solicited from émigrés and military surveyors on exploratory expeditions and the museum situated in Lord Berkeley's town house at 33 Bruton Street in London. On the 11th July 1836 this collection, which included 6720 mammals, reptiles, birds and fishes together with 30,000 insects moved to John Hunter's museum at 11 Castle Street, Leicester Square, London (Desmond & Moore, 1991). These premises had become available by the purchase by the government of Hunter's specimens and their transference to the Royal College of Surgeons in Lincoln's Inn Fields, London.

Hunter built his museum in 1785 on land bought two years previously, and at his death in 1793, it contained more than 2500 specimens including the famous physiological collection of plants and animals. Amongst these was the skeleton of the elephant 'Chunee' which had been exhibited in the Menagerie at The

Exeter Exchange, London. In 1809 this poor animal had become ungovernable and was destroyed by soldiers apparently using 152 musket bullets. The carcass weighed over 5 tons (5000 kg), stood 11 ft (3.3 m) high and was valued dead at £1000. The skin was sold to a tanner for £50 and the skeleton weighing 876lb (400 kg) bought by John Hunter for £100 (Walford, 1987).

The Hunter Collection was housed in already overcrowded facilities at the College of Surgeons and by 1836 political attacks on this establishment, relating to its role as the repository of the nation's unshown treasures, had led to extensive rebuilding. Living on the top floor was the new Hunterian Professor, Richard Owen. Renowned as a leading comparative anatomist, he had been born in Lancaster in 1804 and after some training with local doctors had come to St Bartholomew's Hospital (Bart's), London following time spent in Edinburgh. He served under John Abernathy as Prosector, qualifying MRCS in August in 1826. In 1828 he was appointed lecturer in comparative anatomy at Bart's whilst at the same time employed as assistant curator to the museum of the College of Surgeons. Six years later he was promoted to professor of comparative anatomy at Bart's and in the same year Fellow of the Royal Society (Harrison, 1993). The formation of the Zoological Society of London provided ample opportunity for dissection of unusual species and in 1836 he became Hunterian Professor at the College of Surgeons with responsibility for giving a series of lectures annually on the Hunterian Collection. Until he relinquished this position nineteen years later, Owen gave all of these without ever repeating a topic, a performance never equalled. His audiences were largely 'amateur' representing the great interest shown at this time by the nobility and intellectuals for all matters scientific. The opening of the newly refurbished and enlarged museum in early 1837, attended by the Prime Minister, the Duke of Wellington and more than 500 guests, reflected this interest. Amongst the display were skeletons of chimpanzee, platypus, wombat and many of Darwin's fossilized armadillos and other trophies of his voyages (Figure 1.2). Owen had identified these and by now was the preeminent palaeontologist and comparative anatomist of his day. His paper on the extinct dodo and solitaire remains a classic (Owen, 1879).

The Natural History Department of the British Museum was the national repository for biological specimens but restricted in development by the Trustees whose interests lay in the display of archeological artifacts. Owen was to use his considerable professional reputation and influence to persuade the government to buy 8 acres (19.8 hectares) of land in South Kensington and nine years later build on this site the present Natural History Museum, completed in 1881. He had become Superintendent of the Natural History Department at the British Museum and it was inevitable that he should become the first Director of

Figure 1.2. Sir Richard Owen with the femur and tibia of the Moa.

the new museum for which he had fought for so long. This now housed part of the Hunterian Collection and Museum of Practical Geology thus allowing research at an inter-disciplinary level. In addition to his interest and contributions to palaeontology, Owen had probably dissected more apes than anybody at this time, yet his speech to the British Association for the Advancement of Science in 1854 centred on the 'impossibility of Apes standing erect and being considered in any way related to man'. Initially Owen had a close professional and personal friendship with Darwin. He identified many of the fossils collected during the voyage of the Beagle but his strong Anglican beliefs conflicted with Darwin's propositions in *Descent of Man* and he never came to accept a hypothesis which he believed challenged God's role in Creation. This dissension was never resolved in spite of major contributions to the rapidly developing sciences of zoology, geology, palaeontology and comparative anatomy, and was only touched upon in his highly successful Hunterian Lectures on 'Functions of Animal Organs' (Rupke, 1985).

Owen made fundamental contributions to clarifying classifications of animals and was largely responsible for the reintroduction of the microscope as an essential scientific instrument. By formulating general laws of animal morphology and pioneering the science of palaeontology he encouraged the collection of specimens by travellers from around the world. One of his publications of 1000 pages included details of more than 6000 specimens, and he was the amongst the first to distinguish between the analogous and homologous parts of organisms giving the differentiation a clarity that only his breadth of knowledge made possible. His researches at the College of Surgeons did not however, go unchallenged for he was criticized for using limited finances to study comparative anatomy rather than human morphology or surgery.

London medical schools had included comparative anatomy in their curricula since 1835 and Owen had argued that dissection and animal experimentation greatly added to an understanding of human anatomy and physiology. Indeed, the *Medical Times* of 1844 commented 'No one can acquire a clear insight into the physiology of human organs unless he have borrowed from comparative anatomy the powerful light which that interesting science can alone shed upon his researches' (*Medical Times*, 1844).

By the time Owen resigned his Hunterian Professorship in 1856 every London medical school offered a comprehensive programme of lectures in morbid and comparative anatomy. Comparative anatomy had now acquired the status of an orderly and constructive branch of science with collections embodying living principles and ideas.

Sir Victor Negus

Although Owen described the anatomy of a wide range of animals, including the marsupials, his descriptions were of a generalized nature. Even today, few have the time, funding or motivation to confine their research to studying a single species, let alone a specific region in detail.

Victor Ewings Negus was born in London on the 6th February 1887, receiving his medical education at King's College Hospital where he qualified in 1912. Most of his postgraduate training in diseases of the ear, nose and throat was also in London although travels to the USA did lead to a lifelong interest in the use and design of endoscopic instrumentation. Despite a long and meritorious career in laryngology, it is his pioneer research into the mechanism of the animal larynx which established him as a unique comparative anatomist.

Almost 65 years after Owen's resignation as Hunterian Professor, Victor Negus began work in the museum of the Royal College of Surgeons studying many of Hunter's original specimens (destroyed by enemy action in the Second World War). This material was supplemented by fresh specimens from the Zoological Society of London and during the next eight years he showed that the larynx was a valve whose primary function is to protect the lower (airway) respiratory tract. Voice was a by-product of laryngeal function rather than its prime purpose although important in the ascendancy of humans over other primates. He gave only one Hunterian lecture because by this time these lectures were divided amongst many individuals. In 1929 he published his comprehensive *The Mechanisms of the Larynx* (Negus, 1929). The Preface by Sir Arthur Keith says 'this work shows the same patient power of assembling observation after observation as Darwin had and some of the hot pursuit of function as was urged by Hunter'. Such comparisons cannot be denied when it is appreciated that this research was carried out not by an anatomist or zoologist but by a specialist surgeon still in training (Harrison, 1986). For this work he was awarded the John Hunter medal given triennially by the Royal College of Surgeons of England (1925–1927) and died at the age of 87 years in 1974 still engaged on investigations of the comparative anatomy of the paranasal sinuses.

Studies of comparative anatomy were still in vogue at the beginning of the twentieth century but morphological anatomy was gradually giving way to an increased interest in physiological conceptions. The structural approach to biology was no longer a central preoccupation, although it continued to provide an essential foundation for almost all biological research. When the Zoological Society was founded, doctors and naturalists believed that the structure and functions of all living animals were directed and integrated by vital forces. During the two or three decades which followed publication of Darwin's *On the Origin of the Species* the dominant view changed to a belief that the morphology of an organism was to be understood in terms of the inheritance of a basic structural plan characteristic of its group. This had to be correlated with any specific adaptations which would then distinguish individual species. Morphology was virtually subsumed into evolution, particularly in the primates. Cuvier (1769–1832) asserted that the experimental method was inapplicable to living animals. He maintained that physiological behaviour could be deduced from morphological anatomy, since all parts of the body are interlinked and cannot function unless they act together.

It was inevitable that the intellectual stimulus provided by Darwin's observations should lead to a lessening of emphasis in the study of 'pure' comparative anatomy although there is evidence to suggest that this discipline is now undergoing a renaissance, fuelled by new techniques which clarify the relationship between structure and function. Today, comparative anatomy emphasizes tenable concepts whilst accepting that this discipline cannot answer phylogenetic deductions on its own but can provide reliable guidelines. Reliance on dissections has been replaced by correlation of structure with function in individual species rather than emphasizing similarities. Morphological studies of specific regions relate to an understanding of local function with morphometrics being a better guide to the function of a species as a whole. Grouping variables found in regional anatomical studies makes more sense when considered from the standpoint of function, although these may be modified by heredity.

Much of Victor Negus' work was based on meticulous dissections of a relatively small number of animals, then correlated with the limited knowledge then available regarding laryngeal and respiratory function. New techniques have evolved which allow the larynx to be studied three-dimensionally permitting structures to be correlated with function. In Mammalia, morphometrics allow corporate assessment of structure with quantitative assessment of function. This data can then be used to make comparisons with human performance and subsequent chapters will describe these findings, based on a wide variety of species studied comparatively and with reference to such activities as respiration, locomotion and vocalization.

References

Belon, P. (1551). *L'histoire de la Nature des Oyseaux*. Paris.
Birch, T. (1766). *The History of the Royal Society of London, 1660–1687*.
Camper, P. (1778). Account of the organs of speech of the Orang Outang. *Philosophical Transactions* LXIX. London.

Casserius, J. (1601). *De Vocis Auditusque Organis*. Ferrariae.

Cole, F. S. (1944). *A History of Comparative Anatomy*. London, MacMillan & Co.

Desmond, A. & Moore, J. (1991). *Darwin*. London, Alderman Press.

Dunning, A. (1993). Lessons from the dead. *New Scientist*, 30th January, p. 44–7.

Fabricius, H. (1600). *De Formato Foetu*. Venice.

Gouye, T. (1688). *Observations Physiques Envoyees de Siam a l'Academie Oryal Science*. Paris.

Harrison, D. F. N. (1986). Victor Negus: 57 years later. *Annals of Otology, Rhinology and Laryngology*, **95**, 561–6.

Harrison, D. F. N. (1993). Sir Richard Owen. *Journal of Medical Biography*, **1**, 151–4.

Harvey, W. (1628). *De Motu Cordis et Sanguinis in Animalibus*. Frankfurt.

Hopstock, P. (1921). *Leonardo as Anatomist. In Singers Studies II*. Oxford.

Huxley, E. (1981). *Whipsnade*. London, Collins.

Medical Times (1844). Leader. *Medical Times*, **10**, 350–1.

Negus, V. (1929). *The Mechanism of the Larynx*. London, Heinemann.

Owen, R. (1879). *Memoirs on the Extinct Wingless Birds of New Zealand; with an Appendix on those of England, Australia, Newfoundland, Mauritius and Rodriguez*. 2 vols. London, John van Voorst.

Perrault, C. (1734). *Memoires pour servir a l'Histoire Naturelle des Animaux*. Paris.

Ruini, C. (1598). *Dell' Anatomia et dell'Infirmita del Cavalo*. Bolongna.

Rupke, N. (1985) Richard Owen's Hunterian lectures on comparative anatomy and physiology. *Medical History*, **29**, 237–58.

Severino, M. A. (1645). *Zootomia Democritaea*. Noribergae.

The Charter and Bye-Laws of the Zoological Society of London. (1958) London: Waterlow & Sons Ltd.

Tyson, E. (1680). *Phocaena, or the Anatomy of the Porpoise*. London.

Tyson, E. (1699). Orang-outang, *sive* Homo sylestris: *or the Anatomy of a Pigmie*. London.

Vesalius, A. (1543). *De Humani Corporis Fabrica*. Basileae.

Walford, E. (1987). *Old London – Strand to Soho*. London, Alderman Press.

Willis, T. (1672). *De Anima Brutorum*. Oxford.

2 Collection of specimens and data

In order to confirm the cause of death and absence of infectious disease the majority of animals dying in zoological collections undergo detailed pathological examination. This provides a rich source of material for comparative studies and the majority of the 1410 mammalian larynges studied by the author over the past 18 years have come from the Pathology Department of the Zoological Society of London. Specimens have also been donated by many other zoos, university departments of anatomy and zoology as well as interested individuals from around the world. Some of the more unusual specimens were obtained 'fresh' within the game parks of Kenya and South Africa; only goodwill between many colleagues made this unique collection possible.

Where technically feasible the trachea was removed with the larynx and tongue base, allowing concurrent studies of tracheal morphology and the related recurrent laryngeal nerves. Together with the 1410 non-human larynges, a further 140 laryngectomy specimens were removed from patients with malignant disease (Harrison, 1983) and another 104 larynges removed from infants registered as having died from sudden infant death syndrome (SIDS) plus a control set of larynges from a similarly aged group of 20 infants dying from congenital heart disease (provided by five Paediatric Pathologists within the United Kingdom).

An essential feature has been the need for accurate identification of individual species, together with recording of sex, age and dead weight of each animal, all requiring a high degree of interpersonal expertise.

Classification

Formal classification divides Mammals into two subclasses: Protheria and Theria. These are broken down further. For example, for living Protheria there is only one group, the Monotrema (platypus, spiny anteaters); and for living Theria there are two groups, the Marsupialia (opossums, kangaroos, etc.) and Eutheria (broken down further into orders such as Carnivora, Primates, Insectivora). These divisions are based on morphological characteristics and bear no relationship to physiological behaviour or ecological separation. For example, the 2.5 kg mouse deer (*Tragulus javanicus*) and the 3000 kg hippopotamus (*Hippopotamus amphibius*) belong to the order Artiodactyla (even-toed ungulates) but have little else in common. Additional information based on life history and natural behaviour is used to supplement both anatomical studies and order classification. However, detection and identification of those genetically independent groups that we call species is of considerable importance, although the concept is not rigid or always clear-cut. It is based upon the premise that individuals within a geographically coherent group may show a degree of interspecial resemblance due to a common ancestry. This concept, however, becomes less precise with geographically separated populations. Amongst the antelopes for example, the four principal forms of oryx in Africa and Arabia are all isolated from each other and have been considered races of a single species – *Oryx gazelle* – yet are often treated as three separate species. Corbet and Hill (1991) list 4322 recognized mammalian species, commenting that any new species of ungulates and carnivores are less likely to be identified, whilst several new species of bats and rodents can be expected each year. In identifying the species listed in the Appendices to this book I have utilized the classification system of Corbet and Hill, with the exception that the Scandentia, Dermoptera and Macroscelidea have been absorbed into other orders.

Although the total number of non-human mammalian specimens studied is 1410, this represents only 253 separate species, and there are no representatives of Sirenea (sea cows) or Pholidita (pangolins) because of the lack of any suitable specimens. However, it should be appreciated that within some families such as the Soricidae (shrews) there are 272 recognized species, whilst within the Sciuridea (squirrels) there are 254 species and in the Muridea (e.g. rats and mice) more than 900 species. These numbers may distort the spread of special representation and 253 separate species does represent a reasonably wide coverage of most orders.

The number of babies, juveniles and adults within each species (sexual variations are referred to when relevant in the text) are given in Appendix I. The collection covers a wide range of age groups, although it must be admitted that the definition of juvenile or even infancy can present problems of accuracy in some of the more unusual species. This is particularly important when studying the relationship between structure and function in the young, where all mammals share the same need for ingestion of milk irrespective of their later dietary needs (Pracy, 1984).

Despite the co-operation and support from individuals and organizations within the United Kingdom and elsewhere, availability of unusual species in large numbers is dependent upon practical considerations, such as the popularity of an animal within zoos (e.g. hyaenas are interesting but unpopular), availability within a particular region (e.g. the platypus and giant panda are rare even in their own habitats) and poor access to many regions of the world.

Negus in his pioneering work on the animal larynx (Negus, 1924) described 168 separate species but gave no details of the number of specimens studied. This present larger and more representative collection, analysed with modern techniques and collated with new physiological data, should provide fresh insight into the relationship between structure and laryngeal function.

Collection techniques

Although the larynx, and in smaller mammals pharynx and palate, were removed soon after death and the specimen placed in 10% buffered neutral formalin, delay in collection of the preserved material in some instances took many months. This resulted in some rigidity of the tissues making subsequent macrodissection difficult, although not materially affecting serial sectioning. Technical and geographical difficulties in collecting specimens from sources outside the United Kingdom delayed access to a number of the more unusual larynges. In the case of protected animals such as the polar bear (*Thalarctos maritimus*), considerable documentation was required to ensure trouble-free passage through Customs. In practice no specimen attracted more than a cursory inspection on entry into Britain, perhaps due to the pervading vapour of formalin!

Dissection of material within the game parks and other sites where 'fresh' specimens were obtained was carried out by the author or helpful colleagues; for practical reasons this was limited to the smaller animals. Unfortunately access to 'culling programmes' was frustrated by political sensitivities preventing the collection of large numbers of specimens from a single species. Fortunately, herd animals are common within open zoological parks such as Whipsnade and specimens from 24 zebra (*Equus burchelli*), 17 fallow deer (*Cervus dama*), 19 reindeer (*Rangifer tarandus*) as well as 16 cheetah (*Actinoyx jubatus*) were collected and covered a wide age range.

Documentation

The development of histological techniques for serially sectioning human larynges provided the means for investigating the non-human mammalian larynx three-dimensionally (Tucker, 1963; Harrison, 1970). This requires at least three specimens of each larynx to allow sectioning in the coronal, sagittal and transverse planes, with further information available from macrodissection of additional specimens. Documentation of this data to enable transference to a personal computer (PC) data base for analysis, required a custom-designed punch card which was based upon that developed by the Department of Otolaryngology, University of Toronto, Canada for human laryngeal sectioning (Figure 2.1). Areas were reserved for details of collection data, gross and microscopical anatomy and histological techniques; all being collated under a code number specific to each specimen. This minimized confusion when examining X-rays, histological sections, drawings, photographs or macrodissection details. Batches of punch card data were entered into the data base run on an Apple Macintosh II cx and analysed using FASTAT (SYSTAT Inc.), this equipment replacing the original Amstrad PC 1512 running with STATGRAPHIC software. The punch card designed in Toronto but suitably modified was used for recording the laryngectomy specimens and the SIDS data (Harrison 1976).

Utilization of the specimens

The equipment available for sectioning restricted maximum specimen size to 12 cm × 6.5 cm × 7 cm, which is approximately the size of the larynx of an adult onager (*Equus hemionus*). All larynges smaller than this were set aside for serial sectioning with the aim of acquiring three specimens of each species to enable sectioning to be carried out in the coronal, sagittal and transverse planes. A complete 'set' allowed three-dimensional assessment and with the bat larynges, three-dimensional reconstruction (Denny, 1976). The SIDS larynges were all cut in the transverse plane whilst the majority of the laryngectomy specimens were cut in the coronal plane.

Larynges not destined for processing were displayed, sectioned in either the coronal or sagittal plane as museum specimens. Selection was limited by specimen size and most of the very large and heavy specimens, such as elephant or hippopotamus, remained in formalin-filled containers. The remaining larynges and tracheas were used for macrodissection and the data obtained is discussed in later chapters.

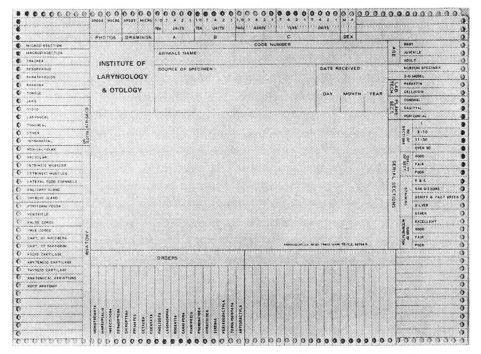

Figure 2.1. Custom-designed punch card for recording data prior to transference to computer data bank.

Prior to the availability of a Reichert-Jung Polycut 'S' automatic microtome, larger larynges were cut using a commercial 'bacon slicer'. This allowed superficial slicing of formalin-fixed specimens and although limited to un-calcified larynges did provide much useful information. This technique has now been adopted by pathologists for the rapid examination of operative specimens of the larynx.

Sectioning technique

The development of methods suitable for serially sectioning malignant larynges (Tucker, 1963; Harrison, 1970) was designed to deal with tissues weakened by neoplastic invasion and used double embedding with celloidin for section stability. This technique was employed for the larger animal specimens such as the horse, but sections are limited to 25 μm in thickness. A modified and quicker version of this technique was used for the majority of the mammalian larynges and those sectioned in the SIDS study (Harrison, 1991).

Following fixation in 10% buffered neutral formalin the whole larynx is X-rayed and then placed in 10% formic acid for decalcification. The end-point is determined after periodic X-ray and can take up to 14 days for a specimen measuring 12 cm × 6 cm × 7 cm. The whole larynx is then processed by a double-embedding technique using low viscosity nitrocellulose (LVN) and wax. This provides adequate support for both soft and hard tissues each of which have varying shrinkage and hardening rates. The time taken to reach the embedding stage varies with each species. After decalcification the larynx is immersed in 70% alcohol for two days, then into three changes of absolute alcohol for periods of two, two and four days. Embedding in 1.5% LVN for between two and three days is followed by chloroform to harden the LVN and also remove all traces of alcohol, which is not miscible with wax. Embedding is in 'fibrowax' which has a melting point of 56 °C and contains plasticizers for maximum tissue support. Three changes are carried out under a vacuum over eight hours and the larynx is then placed in a mould containing fresh wax and allowed to cool.

Microscopy

Small blocks were cut on a manual base-sledge microtome and the remainder with a Reichert-Jung Polycut 'S' automatic machine. This provides the force required to cut large, hard blocks as well as ensuring that the distance between levels remains accurate. Cutting speed and length can be maintained and five sections were taken at each level at a thickness of between five to ten µm, depending on the size and orientation of the specimen. The surface of the block was kept soft by a solution of nine parts of 60% alcohol and one part of glycerine and the wax firmed by an ice tray prior to cutting at each level.

Sections were floated onto 20% alcohol to remove wrinkles, drained and then floated onto a water bath heated to 45 °C. Following positioning on large slides these are placed on a hotplate at 60 °C to melt the wax and allow sections to adhere.

Staining with haematoxylin and eosin, Verhoeff's elastic stain and periodic acid Schiff (PAS) was used in all specimens, with spare sections allowing special staining when necessary.

The plane at which the larynx is cut was decided prior to processing since sagittal sectioning requires only half the larynx, the remaining side being preserved for later study. Collection of three specimens, which permits sectioning to be carried out in the coronal, sagittal and transverse planes, allows three-dimensional evaluation with spare sections available to provide additional data (Figures 2.2, 2.3 and 2.4). All sections were examined with a Wild microscope

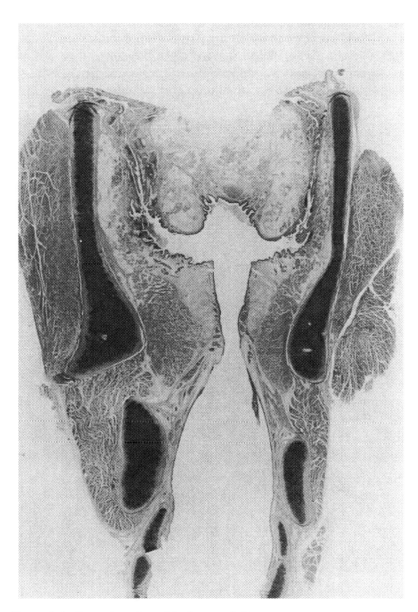

Figure 2.2. Coronal section of the larynx of chimpanzee.

and details relating to anatomy and histology recorded in note form prior to documentation on the punch card. Every section was photographed using an attachment to this microscope and selected sections drawn utilizing a custom-designed side-arm.

In addition to the use of serial sections, a model was developed for the bat larynges. Plastic sheeting was cut into sections which were then attached to the projection face of a Zeiss Ultraphot microscope. Prominent structures on the sections were traced onto the sheet with a Rotring ink pen and the sheets displayed in a specially made aluminium box, allowing three-dimensional

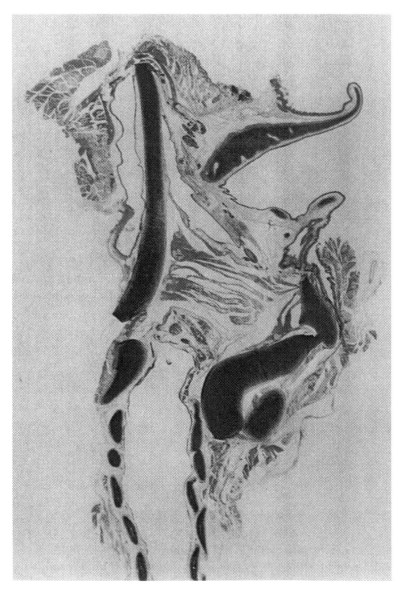

Figure 2.3. Sagital section of the larynx of squirrel monkey (*Saimiri sciureus*).

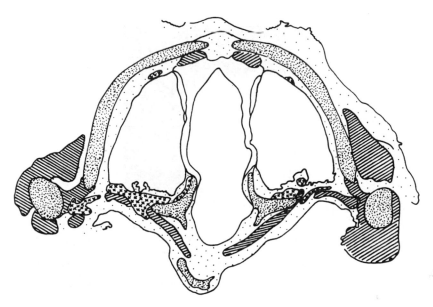

Figure 2.4. Drawing of a transverse section of the larynx of northern night monkey
(*Aotus trivirgatus*).

visualization. The larynges of these species are small and complex and this
method assisted in the evaluation of the microscopical anatomy (Figure 2.5)
Development of image analysing systems with associated statistical software
programmes has allowed accurate measurement of a variety of laryngeal and
tracheal parameters, such as glottic and tracheal area and fibre-size frequency in
the recurrent laryngeal nerves. The data from this is considered in later chapters
particularly with reference to symmorphosis in relation to the relative dimensions of the glottic and tracheal areas, and the significance of fibre-size
frequency in mammalian recurrent laryngeal nerves. This equipment has
proved of considerable value in the measurement of the reduction in the
subglottic airway in infants dying from SIDS, providing a rapid and reproducible alternative to more time-consuming manual techniques.

Macrodissection

Although the use of serial sectioning provided much useful information in larger
specimens where the sections totalled more than 60, interpretation in some of
the smaller specimens proved to be difficult. Features such as the cricoarytenoid
joints and morphology of the laryngeal nerve supply and muscular attachments

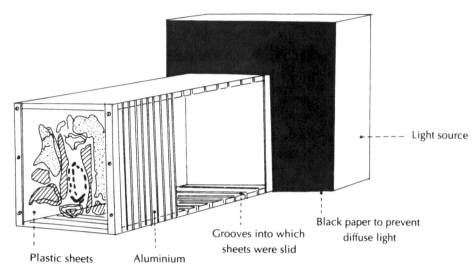

Figure 2.5. Schematic representation of the equipment used to reconstruct bat larynges.

required careful micro- and macrodissection, with meticulous recording of data. In some instances where the number of specimens was limited this replaced sectioning in a plane that was thought probably to be unhelpful, such as the transverse. Special staining techniques to illustrate macroscopic differentiation between squamous and columnar epithelium were used in several groups of mammals to illustrate variations in subglottic hyperplasia. This was correlated with embryological evidence where available, and this relationship between human and non-human laryngeal morphology is a feature throughout this volume.

References

Corbet, G. B. & Hill, J. E. (1991). *A World List of Mammalian Species*. 3rd edn. Oxford University Press.

Denny, S. P. (1976). The bat larynx. In *Scientific Basis of Otolaryngology*, ed. R. Hinchcliffe & D. F. N. Harrison. London, Heineman (Medical Books) Ltd.

Harrison, D. F. N. (1970). Pathology of hypopharyngeal cancer in relation to surgical management. *Journal of Laryngology and Otology*, 84, 349–52.

Harrison, D. F. N. (1976). A long-term comparative study of the mammalian larynx, based on whole organ serial sectioning. *Acta Otolaryngologica*, 81, 167–72.

Harrison, D. F. N. (1983). Correlation between clinical and histological classification of laryngeal cancer. PhD. University of London.

Harrison, D. F. N. (1991). Laryngeal morphology in sudden unexpected death in infants. *Journal of Laryngology and Otology*, **105**, 646–50.

Negus, V. E. (1924). *The Mechanism of the Larynx*. London, Heineman (Medical Books) Ltd.

Pracy, R. (1984). The functional anatomy of the immature mammalian larynx. M. Phil. University of London.

Tucker, G. F. (1963). Some clinical inferences from the study of serial laryngeal sections. *Laryngoscope*, **73**, 728–43.

3 General aspects of laryngeal morphology

It is widely accepted that the larynx evolved around the time that the respiratory system changed from gills to lungs. The earliest larynx consisted of a cartilaginous plate suspended from the vertebral column, as is found in *Protopterus*, the lung fish. With increasing muscular activity, a more dynamic system was needed to provide a wider opening to the lungs and bilateral movable cartilages developed along the sides of the air passage. Sphincteric muscles wrapped around these lateral cartilages provided a means of closing the entrance to the lungs. Later, these fused to form the cricoid ring carrying or articulating with arytenoids on its upper surface. The more developed species formed a protective thyroid cartilage which ultimately articulated independently with the cricoid; this being the basic design of all present day mammalian larynges (Negus, 1949). Although this is so, morphological variations are common, being related in part to life-style and particularly dietary needs.

Most mammals have a well developed epiglottis protruding into the nasopharynx thereby enhancing olfaction and allowing respiration during feeding. The vocal cords vary in length and although arytenoids, corniculate and cuneiform cartilages are usually present, their size and shape depends upon the age and species of the animal. Large arytenoids surmounted by large corniculate cartilages provide a high posterior wall to the larynx, allowing masticated food to flow to the stomach by way of lateral channels on either side of the glottis. This also permits the suprastructure to close over the glottis during regurgitation of rumen, thus avoiding overspill. Such a fold is not present in carnivores who swallow their food without chewing (Figures 3.1 and 3.2). Obviously the form of laryngeal protection depends upon the nature of food eaten and this varies between species. It may be meat or vegetation, swallowed whole or masticated, swallowed–regurgitated–remasticated and swallowed again. Whatever be the nature of the diet eventually it must pass into the digestive system by-passing the entrance to the respiratory system. However, many animals have a mixed diet,

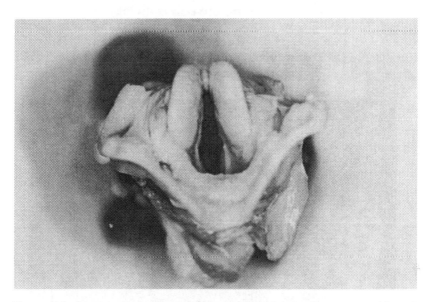

Figure 3.1. Larynx of waterbuck (*Kobus defassa*) showing large arytenoids and corniculate cartilages providing a high posterior and lateral wall to the supraglottis.

for example carnivores eating grass and fruit, whilst some herbivores such as rodents, eat insects. Generally, the higher up the animal kingdom the more diversified the diet and in primates the larynx is a morphological compromise between that found in the specialized herbivores and carnivores. In adult humans the larynx is unprotected by a supraglottis inserted into the nasopharynx and the aryepiglottic fold and arytenoid complex is higher than in other primates, to provide additional protection to the glottis.

Despite those modifications of laryngeal morphology in the adult mammal, which can be related to diet or behaviour patterns, significant differences are also present in infant larynges. Pracy (1984) studied these in detail, considering them to be aligned to the suckling phase of life. Suckling is common to all young mammals and the morphological changes of distension of the aryepiglottic folds by secreting mucous glands are discussed in Chapter 4 (4.4). The associated elevation of the posterior laryngeal wall is designed to protect the lower airway from overspill of milk and, in herbivores, is carried on into adulthood.

General comparative morphology

Herbivores

Plant material is the starting point for all terrestrial animals' food whether they be hunter or hunted. Although abundant, grass has a low calorific value

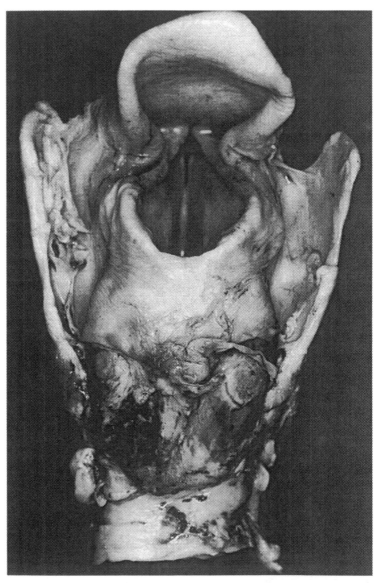

Figure 3.2. Larynx of adult lion showing an absence of large arytenoids which is a feature of carnivores.

requiring the consumption of large quantities and prolonged feeding time. While herbivores feed, the epiglottis and other parts of the supraglottis are held within the nasopharynx by the sphincteric action of the muscles of the posterior pharyngeal wall and soft palate. These enable the animal to breathe, swallow and simultaneously use its olfactory senses for detection of danger. The characteristic morphology is similar in most members of the orders Artiodactyla and Perissodactyla, with an epiglottis which is relatively long, broad and pointed to fit comfortably within the nasopharynx. The aryepiglottic folds provide high lateral laryngeal walls and the large arytenoids narrow the glottic opening. The vocal process of the arytenoids slopes downwards and anteriorly below the level of the upper border of the cricoid cartilage resulting in low vocal cords (Figure 3.3). The herbivore larynx is thus protected anteriorly by a relatively long epiglottis, laterally by high aryepiglottic or epiglottic folds and posteriorly by bulky arytenoid/corniculate cartilage complexes. This circumferential protection is particularly necessary for the ingestion of a purely fluid diet, although in ruminants it is the cellulose remaining in the rumen which is regurgitated. This is not a violent convulsion of stomach contents but a gradual controlled expulsion so the risk of overspill is minimal. In immature ruminants in the suckling stage, the opening into the rumen remains closed and milk passes directly to the final stomach where digestion takes place.

Rodents

The rodent larynx has a rounded epiglottis with the vocal cords sited at the superior border of the cricoid cartilage. The aryepiglottic folds sweep round lateral to the vocal cords in a wider arc than is seen in the Artiodactyla and Perissodactyla.

Carnivores

This larynx is very different from that seen in the grazers, with an epiglottis which appears to 'flap' down over the glottis during deglutition. In the 'cats' there are no aryepiglottic folds and most other species have only epiglottic folds. The arytenoid cartilages are small (Figure 3.2), possibly because of a primary diet of unmasticated food limited in fluid content. Carnivores rarely eat and drink simultaneously, water being ingested by lapping during which the vocal cords are probably closed.

Primates

Characteristically, the primate larynx has an upright rounded epiglottis with aryepiglottic folds that extend from epiglottis to the whole of the vertical face of the arytenoid cartilage. This cartilage is increased in height superiorly by the corniculate cartilage (of Santorini), although variations in height of the posterior laryngeal wall are found within different species.

Although this represents the general morphological characteristics of the

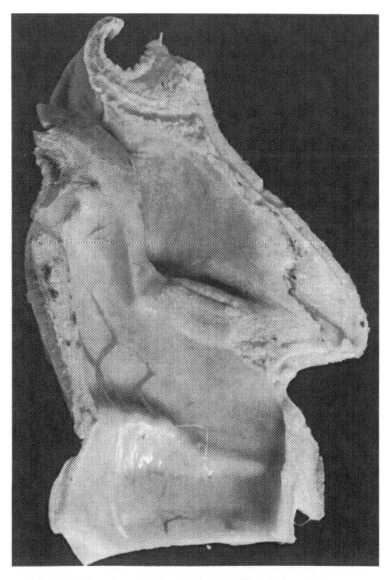

Figure 3.3. Sagittal section of the larynx of a kudu (*Tragelaphus strepsiceros*) showing large arytenoids with vocal process sloping inferiorly below the upper border of the cricoid cartilage.

various mammalian species, individual variations primarily related to diet and behaviour are common. These will be discussed in later chapters with particular reference to specific features such as framework, glandular tissue and musculature. Since structure is thought to be related to function within the animal kingdom, the role of these morphological features will be considered in Chapter 5.

Epithelial distribution

Although Galen in the second century AD, described the membranous lining of the larynx it was not until the nineteenth century that the upper larynx down to the vocal cords was found to be covered by columnar epithelium (Henle, 1838; Koeliker, 1854; Stricker, 1872). Rheiner in 1853 found that the edge of the epiglottis and aryepiglottic fold were covered with squamous epithelium, whereas the remainder of the laryngeal surface of the epiglottis was lined with columnar epithelium. One of the first to investigate the lining epithelium of the larynx using convential histological techniques was Stricker, and in his detailed and accurate account (Stricker, 1872) stated that the anterior surface and upper part of the posterior surface of the epiglottis as well as the vocal cords was lined with squamous epithelium. The remainder of the larynx is covered by ciliated columnar epithelium.

More recent conflicting accounts of the details of the distribution of the epithelium within the larynx have appeared. Hewer (1969) discovered patches of stratified squamous non-keratinizing epithelium within this columnar epithelium. Tucker et al. (1976) studied the whole of the laryngeal epithelium using scanning microscopy and found that the laryngeal surface of the epiglottis was lined by stratified columnar epithelium with non-ciliated and goblet cells widely dispersed throughout the epithelium. The ventricular bands and ventricles were also stratified ciliated columnar epithelium, although the cilia were lost close to the upper surface of the vocal cords. However, convential histological techniques have added little to our knowledge of normal variations in the distribution of both squamous and columnar epithelium within the human larynx. Epstein's 'stripping' procedure (1958) although allowing thorough examination of the entire epithelial field with approximately 300 slides available for study, is both tedious and time consuming. Stell et al. (1972) described a gross staining method which allows rapid assessment of the entire field in in situ. They experimented with two different methods, one utilizing the dye pyronin Y which stains ribonucleic acid. The fixed specimen is immersed in a 4% aqueous solution of pyronin for one minute followed by washing to remove surplus dye.

Squamous epithelium is unstained whilst respiratory epithelium stains a brilliant red. Although rapid and simple to use, the dye diffuses out of the tissue within 24 hours. The technique adopted for their studies of the morphology of the glottis (Stell *et al.*, 1980,) subglottis (Stell *et al.*, 1980) and supraglottis (Stell *et al.*, 1981) used a combination of alcian blue and phloxine (Stell *et al.*, 1972). Alcian blue is an amphoteric dye staining acid mucopolysaccharides; phloxine is a bluish red dye of the xanthene series.

Fixed specimens are immersed in 1% phloxine solution for one minute, washed in running water and then immersed in 1% alcian blue in 5% acetic acid for three to four minutes. Following washing in running water, examination reveals squamous epithelium is stained pink and respiratory epithelium deep blue. The tissues remain coloured in museum specimens for several years and this technique has been used with good effect in the study of human laryngeal epithelium by Stell *et al.* (1980) and in zebra (*Equus burchelli*) and muntjac (*Muntiacus muntjak*) larynges. This method requires reliably fresh specimens that have been washed thoroughly to remove surface mucous and trauma to surface epithelium which is essential to minimize errors in interpretation. Possible sources of error include false positives, due to squamous epithelium staining blue and false negatives, due to non-staining because of surface mucous.

Human supraglottis

The laryngologist defines the supraglottis as consisting of the posterior surface of the suprahyoid epiglottis, infrahyoid epiglottis, aryepiglottic folds, arytenoids, false cords and ventricles. This differs from that adopted by the anatomist who recognizes only individual sites although such disparity is of little practical significance. Stell *et al.* (1981) measured the distribution of squamous and columnar epithelium in 49 human larynges in whom there was confirmed data of an absence of any history of smoking or bronchitis. Distribution between sexes was almost exactly equal and following removal of hyoid bone the larynx was divided into halves by sagittal splits. Following staining by the alcian blue-phyloxine technique the areas of stained epithelium were transferred by pressure to sheets of plastic. Outlines of the areas of respiratory and squamous epithelium were traced on to graph paper for subsequent measurement. This included the total supraglottic area bounded by a plane through the ventricle inferiorly and the epiglottic rim. However, the ventricular epithelium is not accessible for accurate assessment by this approach. The total surface area of an adult human larynx was estimated to be about 22 cm^3 for a man and 15 cm^2 for a woman. The laryngeal surface of the epiglottis was covered by respiratory epithelium with islands of squamous epithelium in 50% of men and women. The posterior surface had a squamous rim occupying about 30% of its surface whilst the remainder of the base was covered by respiratory epithelium. These scattered islands of squamous epithelium appear to be common, although variable in position. During swallowing there is opposition of the epiglottic tip, aryepiglottic folds and arytenoids. This closure is fortified by closure of the false and true cords, all creating areas of possible friction and squamous metaplasia. Although the vestibular folds are generally lined by respiratory epithelium, mixtures with squamous epithelium were found in 40% of larynges and entirely squamous in 10%. Other studies have reached similar conclusions relating the presence of squamous epithelium to ageing.

Human glottis

The nomenclature used to describe the glottic opening is often confusing, with laryngologists referring to the whole area as the vocal cords. This is anatomically incorrect since the glottis contains both vocal cords and arytenoids. The former are formed by the superior free edge of the cricothyroid ligaments which are attached anteriorly to the thyroid cartilage and posteriorly to the vocal process of the arytenoid cartilage. The posterior part of the glottis is bounded by the vocal process and body of the arytenoid and mucosa covering the interarytenoid muscle (Figure 3.4).

Busuttil et al. (1981) examined 93 adult larynges removed within 24 hours of death. They measured the distance between the posterior aspect of the body of the arytenoid cartilage and the tip of the vocal process, and from the tip to the cut mucosa at the anterior commissure.

The lengths of the arytenoids were not related to age but to sex and body height. A linear equation was constructed to reflect this association:

For men the length in inches of the arytenoid = 0.157 + 0.0045 × height.
For women the length = 0.070 + 0.0045 × height.

When differences between the sexes were accounted for, the length of the membranous glottis varied little with either age or height. Mean length for men was 0.486 inches (12.5 mm) and women 0.382 inches (10.6 mm). From these measurements it was clear that 48% of the glottis was composed of rigid cartilaginous arytenoid and 52% the membranous vocal cord. This work was carried out on fixed and decalcified larynges and intrinsic errors secondary to shrinkage are possible. No formal measurements of fixation-induced coefficients of shrinkage have been carried out on the larynx and it is assumed that such errors are of low significance.

Anatomical texts give the lengths of the vocal cords as 15.5 mm in men and 11.5 mm in women, with the intercartilaginous part of the glottis as 7.5 mm in men and 5.5 mm in women. Stell *et al.* (1980) examined glottic morphology in 64 larynges removed from cadavers whose previous history excluded smoking or bronchitis. The vocal cords measured 14.15 mm in men and 10.11 mm in women, with thickness varying from a maximum of 5.5 mm (women 4.23 mm) in the middle of the cord to 4.78 mm (women 2.78 mm) at the anterior commissure. The entire glottis is lined by a band of squamous epithelium being fusiform over the true vocal cords. In most larynges squamous epithelium runs continuously from the anterior end of one cord across the anterior commissure, to the other. In a few cases a narrow strip was present at the anterior commissure which was not lined with squamous epithelium and Tucker *et al.* (1976) using scanning electron microscopy found a strip of modified ciliated cells at this point. The significance of this variation remains uncertain but it is not the route by which mucous clears the larynx (Hilding, 1956).

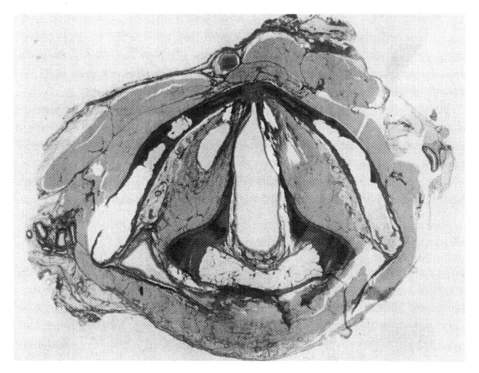

Figure 3.4. Transverse section of a human larynx showing the extent to which the 'glottis' is bounded by the arytenoids.

Human subglottis

Descriptions of the larynx by Henle (1838) and later Tourneux (1885) made no reference to the subglottic space. Even today there is controversy regarding the superior margin for although histologically this is defined as the level at which the glottic squamous epithelium changes to a respiratory form, this cannot be determined by the naked eye. Generally it is agreed that the lateral limits of the subglottic space are the cricoid cartilage and conus elasticus with the inferior margin the inferior border of the cricoid cartilage. The cricothyroid membrane is important as a barrier to the spread of primary subglottic neoplasms and Caprosa (1957) found it to be 2.9 cm^2 in area with an average height in the midline of 0.90 cm (range 0.50–1.20 cm). Carter & Myers (1979) examining the subglottic space in 50 adult larynges, found little variation with age of the cricothyroid membrane in the anterior midline. The mean distance for men was 8.1 mm and for women 6.9 mm. However, all measurements have so far confirmed that women have significantly smaller subglottic dimensions than men. In several of the female larynges the distance from anterior commissure to inferior border of thyroid cartilage was no more than 7 mm, this could be of significance in the extralaryngeal spread of subglottic cancer (Harrison, 1971).

Heymann (1889) and Kanthanck (1890) described islands of squamous epithelium within the subglottic space believing them to be of embryological origin or the result of metaplasia. Little interest was shown in this topic until Hopp in 1955 studied the development of the laryngeal epithelium. He also found scattered areas of squamous epithelium, considering that they might assist in the understanding of the development of laryngeal cancer. Stell *et al.* (1980) as part of their study of the morphology of the human larynx examined the subglottic mucosa of 63 larynges by means of surface staining. Total extent of the subglottic space was 1297 ± 36.71 mm^2 in men and 879 ± 19.93 mm^2 in women. Over half of the larynges showed some squamous epithelium although this was not correlated with age or sex. Yung *et al.* (1984) using a similar method examined the subglottic area in 34 larynges taken from stillborn babies and infants dying aged less than three months. Histological confirmation that areas staining pink with alcian blue-phloxine were squamous epithelium was carried out in all cases. Three of the 11 stillborn children had areas of squamous epithelium in the subglottis, three of the eight babies with no history of previous intubation also showed some squamous epithelium. Almost half of the babies who had been intubated had subglottic squamous epithelium. The presence of squamous epithelium within the larynx without any history of previous trauma suggests either a congenital origin or possibly postnatal infection. However, in the stillborn, infection is not a factor supporting a congenital origin.

Human foetal larynx

The human larynx is exposed to a variety of traumas such as vocalization, dusts and smoking. In the foetus, the larynx is free from such irritants and studies of the laryngeal epithelium provide a baseline for comparisons with adult morphology. Tourneux (1885) was the first to study the foetal epithelium but subsequent research has been hampered by difficulties in obtaining adequate material. Stafford and Davies (1988) removed larynges from ten aborted mid-trimester human foetuses of between 14 and 18 weeks gestational age (as assessed by ultrasound scanning and the date of the last menstrual period). These abortions were not for foetal abnormalities and termination was induced with prostaglandin E, which unlike hypertonic urea does not cause epithelial damage. Following fixation in 2.5% glutaraldehyde within 40 minutes of termination the specimens were sectioned along the midline sagittal plane. The 20 hemilarynges studied by scanning electron microscopy all showed the same general pattern of epithelial distribution with some variation in the ratio between squamous and columnar epithelium.

Tourneux (1885) had observed that the vocal cords were covered by squamous epithelium and this is apparent by 14 weeks' gestational age. However Stafford and Davies also found scattered areas of ciliated cells along the entire length of the vocal cords in eight out of ten larynges. They concluded that this suggested a close developmental association between these two types of epithelium, and at the anterior commissure two larynges showed a narrow vertical band of ciliated cells.

The lingual surface of the epiglottis and free edge of the laryngeal inlet were lined with squamous epithelia, which was closely, associated with respiratory epithelium. Whilst the remainder of the supraglottic region was covered by ciliated mucosa, occasional islands of squamous cells were found throughout in eight of the larynges. Stell et al. (1981) had also found this in more than 50% of the larynges of non-smoking adults, although Hopp (1955) considered that these islands resulted from metaplasia. This would now appear inappropriate in the absence of a history of trauma, infection or smoking, and their presence in stillborn infants or babies without a history of intubation suggests a congenital origin.

Ageing in the vocal cords and muscles

Clinical evidence of changes in the voice with age have been reported by many speech researchers including Shipp & Hollien (1969), Bohme & Hecker (1970) and Honjo & Isshiki (1980). This can be attributed in part to structural changes within the vocal cord and this has been investigated in 64 Japanese larynges, aged between 70 and 104 years, by Hirano *et al.* (1989). The thickness of each layer of the mucosa was measured in the middle of the membranous vocal fold. Histological changes were focused on the superficial layer of the lamina propia, density and possible atrophy of elastic fibres in the intermediate layer of the lamina propia as well as fibrotic changes in the collagenous fibres deep within this layer. The larynx of a 27-year-old non-smoking male provided baseline data using a 6-point scale: zero equalled no abnormality and 5 maximum deviation from normal. Evaluation was carried out on the mid-section of the membranous vocal cord.

While it was accepted that considerable individual differences were to be found in histological changes within the vocal cords, Hirano *et al.* (1989) concluded that:

A. As age advances in males the membranous vocal cord shortens, the elastic fibres in the intermediate layer of the lamina propia atrophy and the collagenous fibres in the deep layer become fibrotic.

B. After the age of 70 years the thickness of the covering epithelium increases and these changes cause an increased stiffness of the vibrating tissues with an increase in fundamental frequency. In females this shortening of the membranous vocal cord is less marked as are the fibrotic changes found within the lamina propia. Only slight changes in fundamental frequency are to be expected but the increased thickness of the covering epithelium, which is a feature of the ageing process, itself leads to a decrease in fundamental frequency.

Their conclusions differed from those of Rodeno *et al.* (1993) in their histo-chemical and morphometric study of the thyroarytenoid and posterior cricoarytenoid muscles and vocal cords in 43 human larynges (aged between 46 and 87 years). Ageing changes consisted of a low increase in the percentage of type 1 and decrease in type 2 muscle fibres in thyroarytenoid; decrease in type 1 fibre percentage and size with an increase in type 2 fibres in the posterior cricoarytenoid muscle. In addition they found a decrease in the thickness of the mucosa and lamina propia of the vocal cords and less perimysial tissue. There was no statistically significant change in the density of subepithelial mucous glands in the vocal cords.

However, these studies emphasize variations in histological changes which may occur in the laryngeal part of the ageing process, and where environmental and life-style factors can play such an important role.

Comparative morphology

Superficial examination suggests that the gross morphology of the mammalian larynx follows a basic pattern, particularly in the immature animal. The greatest variation is seen in the carnivores where there is absence of any need to swallow large quantities of liquid with the bolus of food. The arytenoids are relatively low with no lateral aryepiglottic folds to shield the laryngeal inlet from overspill. All immature animals require laryngeal protection from their predominantly liquid diet; the means by which this is achieved and the modifications which occur at maturity are discussed in later chapters. Indeed, the most obvious differences which are to be found between adult and infant larynges are in the lateral region of the larynx and are primarily related to dietary intake. In herbivores such as camel (*Camelus* spp) and onager (*Equus hemionus*) the side walls extend as high as the upper border of the epiglottis, the functional result being that the lateral laryngeal walls lie within the epipharynx held by the nasopharyngeal musculature. Overspill is virtually impossible.

The specialization of the respiratory system of cetaceans reflects their response to the adoption of a permanent aquatic life-style. Interspecies variations occur in the passage from the blow-hole to the lungs where it crosses the pharyngeal part of the alimentary canal, as in all mammals. Protection of the airway is obtained by the enlarged epiglottis and arytenoids forming an elongated spout inserted into the posterior aspect of the nasal cavity, held firmly in position by the palatal musculature (Burns, 1952; Figure 3.5).

Although this feature of the gross morphology of the cetacean larynx was first recognized by Belon in 1551 (see Chapter 1) the presence of vocal cords and their role in sound production has been disputed as recently as 1981 (Mackay & Liaw, 1987). Reidenberg & Laitman (1988) investigated the larynx in 24 odontocetes representing 10 genera (*Delphinus, Stenella, Lagenorhynchus, Tursiops, Grampus, Delphinapterus, Globicephala, Kogia, Mesoplodon* and *Phocoena*). Contrary to 'established belief' vocal folds were present in all the specimens examined. They were not isolated bands but continuous with the internal laryngeal membrane with attachments to arytenoid and thyroid cartilage as in other mammals. Some variation in morphology was present in that the folds were bifurcated, trifurcated or even a single midline structure. Ventricles and vestibular folds were present lateral to the cords in all the specimens. The larynges came from all ages and ranged from foetuses (two), infants, juveniles and adults although individual numbers were small. Most were found dead on beaches or in nets and the age was determined by criteria such as: foetus – umbilical cord attached; infant – fringed tongue, non-erupted teeth and milk

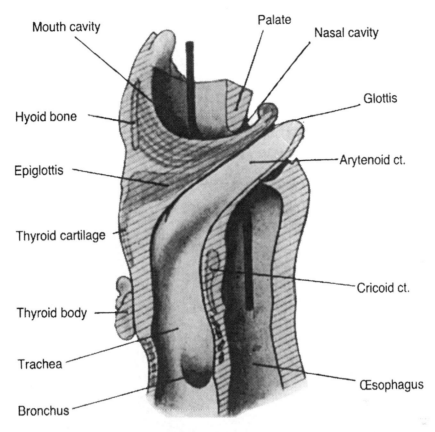

Figure 3.5. Diagramatic illustration of a sagittal section through the laryngeal region of a dolphin (*Delphinus delphis*). (After Burnes, 1952.)

diet; juvenile – small size, erupting teeth. Even foetal and infant specimens exhibited well-formed midline and lateral folds which were thinner than those found in adults. The caudal attachment to the base of the arytenoid varied from a distinct cartilaginous process, to a separate medially situated projecting cartilage articulating with the arytenoid base. Preliminary histological examination of the vocal folds show them to contain collagen and elastic tissue which is part of the fibroelastic layer, homologous with the cricothyroid membrane in terrestrial mammals. The role of these structures in the production of 'whale sound' is discussed in Chapter 5 (5.3).

Another group of highly specialized echolocaters, the Chiroptera, also have 'basic' mammalian larynges. Depending upon the species, sound is produced through mouth or nostrils but the epiglottis is in contact with the palate or within the nasopharynx in all bats. The ossified cartilages act as a rigid scaffold for the attachment of the unusually large intrinsic musculature that is necessary for sonic and ultrasonic emissions (discussed in Chapter 4 (4.2)).

The vocal cords are thin and relatively short but a need to vocalize and breath simultaneously in flight has resulted in glottic modifications. The region of the posterior commissure shows a substantial expansion through which respiration can occur, even when the cords are adducted for vocalization (Figure 3.6). Large tracheal air sacs are found in possibly all constant frequency pulse emitters, although their role as resonators remains unproven (Denny, 1976).

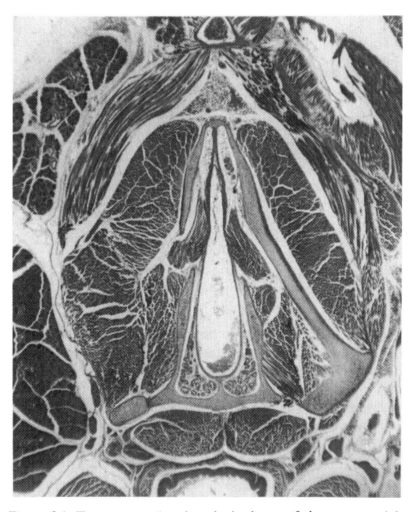

Figure 3.6. Transverse section through the larynx of the spear-nosed bat (*Phyllostomus hastatus*) showing airway expansion posterior to vocal cords.

It has not been possible to carry out detailed studies on the lining of enough non-human species to warrant more than a few superficial observations. Alcian blue-phloxine staining was used on hemilarynges from 12 zebra aged from eight months to adulthood and 6 Indian muntjac aged from two years to adulthood. All had been bred in captivity in urban zoos. Despite prolonged fixation differential staining presented no difficulty and although detailed measurements of the areas of squamous and respiratory epithelium were not carried out, distribution bore close similarity to that reported for humans. Scattered areas of 'stained' squamous epithelium were present in all subglottic regions and confirmed histologically. The specimens have been preserved with staining intact for later detailed examination.

Animals with specialized dietary needs such as the Myrmecophagidae (American anteaters), might be expected to have developed additional defences for their respiratory system. The one specimen examined of the giant anteater (*Myrmecophaga tridactyla*) shows surface epithelium to be covered with short vertical hairs which may act as a barrier to the entry of live ants to the lower respiratory tract.

Comparative studies of the immature larynges of a wide variety of mammals suggest that morphology is primarily related to dietary needs (i.e. ingestion of milk) rather than a need for speedy locomotion or vocalization. The adult larynx has undergone modifications suited to the animals' life-style and these will be examined in detail in subsequent chapters. Not all variations in morphology, however, can be explained on cogent or physiological grounds and these will be discussed in Chapter 6.

References

Bohme, G. & Hecker, G. (1970). Gerontologische Utersuchungen uber Stimmumfang und Sprechtstimmiage. *Folia Phoniatria*, 22, 176–84.

Burns, R. H. (1952). *Handbook of Cetacean Dissections*. London: British Museum (Natural History).

Busuttil, A., Davis, B. C. & Maran, A. G. D. (1981). The soft tissue/ cartilage relationship in the laryngeal glottis. *Journal of Laryngology and Otology*, 95, 385–91.

Caprosa, T. (1957). Practical aspects of the cricothyroid space. *Laryngoscope*, 67, 577–82.

Carter, D. R. & Meyers, A. D. (1979). Anatomy of the subglottis. *Otolaryngology Head and Neck Surgery*, 87, 203–6.

Denny, S. P. (1976) The bat larynx. In *Scientific Basis of Otolaryngology*, ed. R. Hinchcliffe and D. F. N. Harrison, pp. 346–70. London, Heinemann (Medical Books) Ltd.

Epstein, E. E. (1958). A 'stripping' technique for the examination of the total epithelial surface of the larynx. *Journal of Pathology and Bacteriology*, 75, 472–80.

Galen, C. (1823 edition) In Opera Omnia, Vol III, p. 525, 551, 555, 586. C. G. Kuhn. Leipsig.

Harrison, D. F. N. (1971). The pathology and management of subglottic carcinoma. *Annals of Otology, Rhinology and Laryngology*, 80, 6–12.

Henle, J. (1838). Uebec die Aushreitung des epithelium im Menschiehen Koerper. *Archives of Anatomy and Physiology*, 76, 103–28.

Hewer, E. R. (1969) *Textbook of Histology for Medical Students*. 9th edn. London, William Heinemann.

Heymann, R. (1889). Beitrag zur kenntniss des Epithels und der Druesen der Menschlichen Kehlkopfes. *Archives fur Pathologische Anatomie*, 118, 320–48.

Hilding, A. C. (1956). On cigarette smoking, bronchial carcinoma and cilary action. *Annals of Otology, Rhinolgy and Laryngology*, 65, 736–40.

Hirano, M., Kurita, S. & Sakaguchi, S. (1989). Ageing of the vibrating tissue of human vocal cords. *Acta Otolaryngologica* (Stock.), 107, 428–33.

Honjo, I. & Isshiki, N. (1980). Laryngoscopic and voice characteristics of aged persons. *Archives of Otolaryngology*, 160, 149–50.

Hopp, E. S. (1955). The development of the epithelium of the larynx. *Laryngoscope*, 65, 475–99.

Kanthack, A. A. (1890). Studien ueber die histologie der Larynx-Schleimhaut. *Archives fur Pathologische Anatomie*, 120, 273–94.

Koeliker, A. (1854). In *Manual of Human Histology*, vol 2, p. 162. London, Sydenham Society.

Mackay, R. S. & Liaw, H. M. (1987). Dolphin vocalisation mechanisms. *Science*, 212, 676–8.

Negus, V. E. (1949). *The Comparative Anatomy and Physiology of the Larynx*. London, William Heinemann.

Pracy, R. (1984). The functional anatomy of the immature mammalian larynx. Thesis accepted for the Degree of MPhil. University of London.

Reidenberg, J. & Laitman, J. T. (1988). Existence of vocal folds in the larynx of Odontoceti (toothed whales). *Anatomical Record*, 221, 884–91.

Rheiner, H. (1853) Ueber den Ulceration prozen im Kehikopf. *Archives fur Pathological Anatomy und Physiologie*, 5, 534–79.

Rodeno, M. T., Sanchez-Fernandez, J. M. & Riuera-Pomar, J. M. (1993). Histochemical and morphometric ageing changes in human vocal cord muscles. *Acta Otolaryngologica* (Stock.), 113, 445–9.

Shipp, T. & Hollien, H. (1969). Perception of the ageing male voice. *Journal of Speech and Hearing Research*, 12, 703–10.

Stafford, N. D. & Davies, S. J. (1988). Epithelial distribution in the human foetal larynx. *Annals of Otology, Rhinology and Laryngology*, 97, 302–7.

Stell, P. M., Gregory, I. & Watt, J. (1972). Technique for demonstrating the epithelial lining of the larynx. *Journal of Laryngology and Otology*, 86, 589–94.

Stell, P. M., Gregory, I. & Watt, J. (1980). Morphology of the human larynx. 1 The subglottis. *Clinical Otolaryngology*, 5, 389–93.

Stell, P. M., Gudren, R. & Watt, J. (1981). Morphology of the human larynx. III The supraglottis. *Clinical Otolaryngology*, 6, 389–93.

Stricker, A. (1872). *Manual of Human and Comparative Histology*. London, New Sydenham Society.

Tourneux, M. F. (1885). Sur le development de l'epithelium et des glands du larynx et de la trachea chez l'homme. *Memoires de la Societe de Biologie.* (8th series), 12, 250–2.

Tucker, J. A., Vidic, B. & Tucker, J. F. (1976). Survey of the development of laryngeal epithelium. *Annals of Otology, Rhinology and Laryngology*, 85, (Suppl. 30)

Yung, M. W., Barr, G. & Stell, P. M. (1984) Squamous epithelium in the subglottic region of paediatric larynges. *Clinical Otolaryngology*, 9, 145–7.

4 Detailed morphology

4.1 Development of the larynx

For many years embryologists have avoided a mechanistic analysis of cranio-facio-cervical morphogenesis, largely because of the three-dimensional complexity of these regions. Despite its anatomical intricacy, development is more profitably explained in molecular or genetic terms and the genome is often analogized as a genetic blueprint. Genes involved in embryogenesis, however, more accurately serve as a set of assembly rules whose implementation and interaction govern the transition from one-dimensional information of the genotype into the three-dimensional complexity of the phenotype (Thorogood & Ferreti, 1992). The analysis of the development of the embryo is often achieved by the use of model systems or is extrapolated from the study of non-human vertebrates. Since evolution is a conservative process retaining and refining efficient morphogenetic processes, causal mechanisms are conserved over many years simply because they generate a successful phenotype. This appears to be true for the mammalian larynx where much of our knowledge has come from a detailed study of the human embryo. Although the development of the human larynx was described in 1820 by Fleischmann (cited by Soulie & Bardier, 1907) it was not until the pioneer work of His in 1885 on the development of the gastrointestinal tract and pharyngeal arches that serious attention was paid to the larynx.

The traditional method, probably first described by Born in 1883 and used by many early embryologists, consisted of copying microscopical sections after magnification on to wax plates. The wax is then removed, except from those areas to be reconstructed. Alternatively, drawings or photographs from the sections on cardboard or plastic can be cut out and reassembled as used by Denny for the bat larynges and described in Chapter 2. Wind (1970) has described his technique using both paper patterns and polymethylmethacrylate plates to produce spatial reconstructions of very early human embryos. This requires that the angle between the long axis of the embryo and the plane of cross-section be taken into account to avoid distortion in the final reconstruction. Similar reconstructions have been carried out by O'Rahilly & Tucker (1973) on human embryos of the first five weeks (to stage 15) and Müller *et al.* (1985) in eight serially sectioned embryos of stage 23. Other workers, such as Zaw-Tun & Burdi (1985) used combinations of light microscopy and wax-plate reconstructions.

Staging and nomenclature

O'Rahilly & Tucker (1973) emphasized the disadvantages inherent in the arrangement of embryos in series based on previously used criteria. Individuals cannot be arranged perfectly because of variations in the timing of development; it may also prove impossible to match any embryo exactly with any one of the 'standard' norms. A more flexible system of embryonic staging was introduced to minimize these potential errors. In the Carnegie system (O'Rahilly, 1973) human embryos are arranged in 23 groups that cover the embryonic period proper, that is the first eight postovulatory weeks of development. These stages are based on morphological criteria being independent of such variable parameters as length and presumed age. This system therefore allows much greater precision in the establishment of the sequence and timing of developmental events. At the end of the embryonic period proper (stage 23) the human embryo will be approximately 30 mm in crown-rump (C–R) length and eight weeks of age (Table 4.1). The visceral pouches of embryonic reptiles, birds and mammals bear little resemblance to the gill slits of adult fish; rather they resemble the visceral pouches which appear in the embryonic stages of the fish. It is more accurate to say that the fish preserves and elaborates its visceral pouches as gill slits, whilst reptiles, birds and mammals convert them into other structures. O'Rahilly & Tucker (1973) recommended that the term 'branchial' be replaced by 'pharyngeal' from mammalian embryology defining pharyngeal arch as: 'One of a series of lateral, mesenchymal elevations that extend ventrally as arches around the developing pharynx'. A pharyngeal pouch was then defined as 'One of the lateral expansions of the cavity of the developing pharynx, which intervene between the pharyngeal arches'. The assignment of various laryngeal structures to specific pharyngeal arches, e.g., the epiglottis to the fourth and perhaps third arch, remains in doubt. Even the number of pharyngeal arches in the human embryo is a matter of uncertainty, although four arches and pouches are recognized by stage 13. The difficulties in tracing pharyngeal derivatives are seen even in the first pouch. Although the eustachian tube and tympanic cavity are said to be derived from this pouch, it has been suggested that it disappears

Table 4.1. *The Carnegie system of staging*

Carnegie stage	Usual lengths (aver.)	Average age (days)
	1.0–1.5	18
9	1.5–2.5	20
10	2.0–3.5	22
11	2.5–4.5	24
12	3.0–5.0	26
13	4.0–6.0	28
14	5.0–7.0	32
15	7.0–9.0	33

Source: O'Rahilly & Tucker, 1973.

completely in the early embryo before the middle ear appears (Goedbloed, 1960).

Negus (1924) believed that the respiratory apparatus developed as a ventral outpouching from the foregut, the entrance then becoming the glottis. His diagrams illustrating the evolution of the larynx show a muscular sphincter surrounding the glottis in which area the surrounding mesoderm differentiates into cartilage and muscle to form the larynx. Development of the respiratory apparatus in the human is similar to that found in other mammals, with the respiratory primordium budding off the primitive foregut and growing caudally on a lengthening stalk. The cephalic end of this stalk forms the glottis and infraglottis whilst the remainder becomes the trachea. This laryngotracheal sulcus begins to become circumscribed about stage 10 and expansion of the extreme caudal end represents the unpaired, symmetrical pulmonary primordium. Following the appearance of the tracheoesophageal septum at stage 12 and the hypopharyngeal eminence, arytenoid swellings and epithelial lamina at stages 14 and 15, the larynx becomes established as a definitive organ (O'Rahilly & Tucker, 1973).

During these early stages of development part of the laryngeal cavity becomes temporarily obliterated separating the pharyngeal cavity from the trachea. Controversy still exists as to whether this obliteration occurs cephalic to the tracheal opening, as a result of fusion of the arytenoid swellings, or because of compression of the pharyngeal mesoderm. Further confusion is introduced by studies on the rat embryo by Walander (1950) who sited the level of the glottic opening within the epithelial lamina.

Shikinami in 1926 said 'if one wishes to become acquainted with the structure and form of any organ, there is perhaps no better way than to trace its development, step by step, back into the early embryonic stages'. Although the development of the larynx in the human embryo has been studied by many, interpretation is dependent on availability of large numbers of accurately staged specimens such as the Carnegie and Patten Collections. Reconstructions carried out by O'Rahilly & Tucker (1973), Müller *et al.* (1985) and Zaw-Tun & Burdi (1985) allowed a three-dimensional evaluation but even so it was not found possible to reach a consensus on the interpretation of all stages. By stage 23 some indications of the definite shape of the laryngeal skeleton can be recognized, with the future epiglottis seen as dense mesenchyme and deeply staining condensation of precartilage cells marking the site of other cartilages. An important characteristic of this stage is the separation of the epithelial layer between arytenoid swellings, allowing communication between pharynx and infraglottic cavity. Topographically the larynx occupies a more cranial position than later with the epiglottis almost at the level of the upper border of the palate and the arytenoids well above the hyoid body. The epiglottis and palate are not in contact because, at that stage, the lateral processes of the palate have not fused caudally. Even in early foetuses (41 to 43 mm) the region of the soft palate remains partially open and complete closure does not occur until the foetus is between 47 and 52 mm (Müller *et al.*, 1985).

Zaw-Tun & Burdi (1985) describe the sites of chondrification within the developing laryngeal cartilages noting four centres in the thyroid at stage 19. Two on each side, one for the dorsal margin near its cephalic end and the other for the major part of the lamina. In two specimens they noted a gap between the fusion of the caudal part of the dorsal chondrification and the dorsal border of the chondrifying lamina. A small neurovascular bundle passed through this gap which in one of the specimens had become a foramen. This is possibly the origin of the foramen thyroideum found in many mammals (see section 4.2). The disruption of the epithelial layer in stage 23 is an active process rather than simple autolysis and is followed by formation of ventricles as solid outgrowths above the level of the future glottis. Musculature is recognizable as undifferentiated mesenchyme close to sites of future cartilages; the laryngeal lining mainly consists of non-ciliated cuboidal epithelium in two or three layers (Wind, 1970). By the 12th week the embryo (about 47 mm) still shows typical foetal features although its general morphology is approaching that found at birth. The laryngeal lumen is continuous with the trachea through a small opening at the level of the large arytenoids. The skeleton consists of pre-cartilaginous tissue although the two halves of the thyroid are now fused anteriorly. Although the

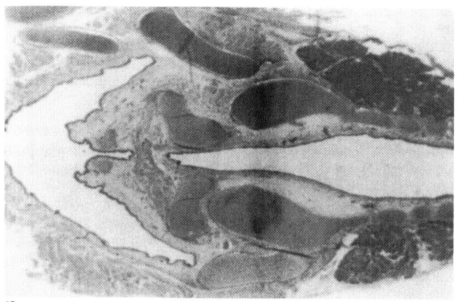

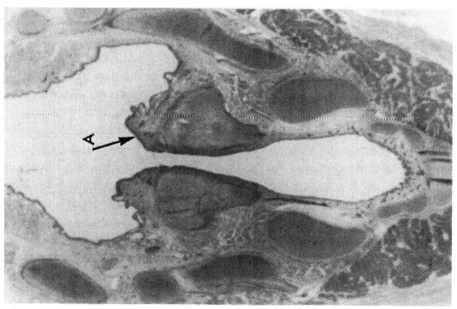

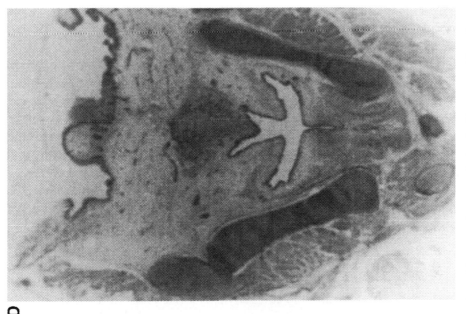

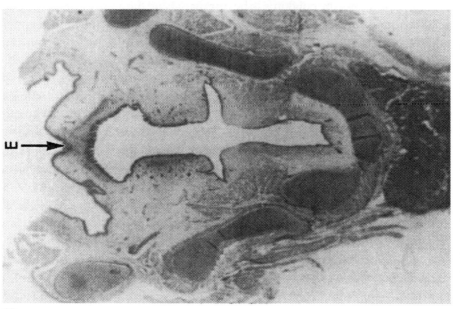

Figure 4.1. Coronal sections from a 24th week human foetus commencing with a posterior cut at the level of the arytenoid cartilages (A) as shown in figure 4.1A, and extending anteriorly to the anterior commissure. Figure 4.1C, shows the early epiglottis (E).

arytenoids articulate with the cranial portion of the cricoid, muscular and vocal processes are ill-formed as are the corniculate and cuneiform cartilages. Most of the intrinsic muscles are recognizable.

There is a gradual maturation of all laryngeal features until birth, although the glottis retains a T-shape as a result of relative retardation in growth of the arytenoids. The epiglottis only shows chondrification at about five months and the thyroid looses its cartilaginous connection with the superior cornu of the hyoid. At this time the arytenoids reach their maximum relative size with the appearance of vocal and muscular processes. All intrinsic musculature is discernible, with the interarytenoid and thyroarytenoids forming a recogniz-able sphincter. Ventricles and saccules are relatively larger than in the adult although the vocal folds are both shorter and blunter than seen postnatally (Figures 4.1A–4.1D).

Position in the neck

In all terrestrial mammals the adult larynx is located high in the neck, usually extending from basiocciput or first cervical vertebra to C3 or C4 (Figure 4.2). This position allows the epiglottis to be intranarial and the animal to breathe and swallow liquids simultaneously. The postnatal development of the human upper respiratory tract is unique among mammals. Infants have a larynx positioned high in the neck as in other mammals. This lasts until the second year of life when the larynx begins its slow descent towards the adult position. There is therefore a separation of the epiglottis from the palate and, briefly, a common pathway for respiration and deglutition between the caudal end of the palate and the upper sphincter of the oesophagus. Inconsistencies and confusion exists, however, regarding the position of the foetal larynx and whether it is intranarial. Variability among reported data is due to measurements taken on foetal heads in different degrees of extension or flexion, inadequate sample size, shrinkage of preserved specimens and lack of standardization for dating the age of foetal specimens by external measurements (Magriples & Laitman, 1987). Although it is accepted that the epiglottic cartilage appears in the fifth foetal month, mesenchymal cells have been seen in this region as early as the fourth month, although not making contact with the palate. Magriples & Laitman (1987) studied mid-sagittal sections of 30 foetuses aged between 15 to 29 weeks with a C–R length of 12.4 to 25.5 cm. The epiglottic cartilage was present at 15 weeks and by 21 weeks was well developed and in close palatal apposition. Full contact occurred from 23 to 25 weeks. Although these studies are of interest, post-mortem examination does not allow precise functional interpretation. Full

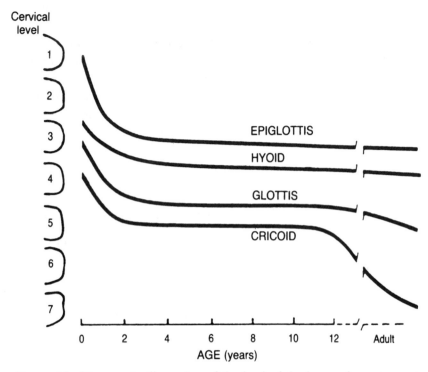

Figure 4.2. Diagramatic illustration of the level of the laryngeal structures at different ages in relation to the cervical spine. (After Westhorpe, 1987.)

clarification of positional relationships require observation of foetal upper airways within natural settings. Wolfson & Laitman (1990) studied the upper respiratory and digestive tract. They found that major laryngeal structures could be seen clearly during normal speed and repeat frame-by-frame analysis in 80% of foetuses. These were defined by anatomical shape, location and high echogenicity. Well-defined thyroid cartilage was identified in 36% of subjects as a large, clearly defined echogenetic structure in the anterior neck. The cricoid was seen in 32% of cases and the epiglottis in 16%; none of the smaller cartilages, arytenoid, corniculate or cuneiform were seen clearly. Since laryngeal movements on swallowing were poorly visualized this technique has little value in physiological assessment.

The child's larynx is narrower and shorter than the adult and the epiglottis more U-shaped. Although recognized as being higher in the neck the position in relation to the cervical spine changes with flexion and extension, opening and closing of the mouth and changing from supine to the erect position. Westhorpe

(1987) studied the relationship of the epiglottis, hyoid and glottis to the cervical spine during normal growth in children from birth to puberty. Measurements were made on standardized lateral radiographs of 30 children aged between one day and 12 years and a further 20 infants aged between one day and three years. A standardized position was used to eliminate errors due to positional variations. The data confirmed that the larynx and hyoid descended in the neck between birth and three years. Little change then occurred until puberty when there was further descent due to growth of the thyroid cartilage. Recognition of these changes was of practical importance to the paediatric anaesthetist in positioning the head and neck for easy, safe intubation.

Effect of basicranial flexion on larynx and hyoid position in rats

Changes in the shape of the basicranium affect the anatomy and function of the larynx and hyoid. Surgically induced flexion in 13-day-old rats (*Rattus norvegicus*) showed that, by ablation of the spheno-occipital synchondrosis, the larynx and hyoid were displaced inferiorly (Reidenberg & Laitman, 1991). This confirms mechanical interaction between the skeletal and soft tissues.

The skull base serves as attachment for the muscles, ligaments and connective tissue which suspend the pharynx, hyoid and larynx and associated structures. Evolutionary changes in shape in adult humans will affect the position of the larynx and size of the pharynx, indirectly influencing acquisition of speech. Analysis of basicranial shape in paleolithic skulls has provided some explanation for the dating of human speech acquisition and is discussed elsewhere. Such studies have been based on morphological observations but may now be augmented by data from experimentally produced variations in skull base relationships.

4.2 Framework, cartilages, ligaments and related joints

Comparative observations of modern amphibians suggest that in early terrestrial vertebrates the breathing of air was added to gill and skin respiration. To achieve this the larynx in amphibians such as the Salamander, shows a differentiation from the primitive protective sphincter found in the lungfish by the addition of lateral and cricotracheal cartilages. In Anuran amphibians such as the frog, the larynx consists of paired arytenoid cartilages with a single circular cricoid cartilage and a well developed muscular system. Evolution of the first true mammals, probably as small insectivores, necessitated a more complex organ to modulate respiratory protection, control of airflow and later, sound production. This was achieved by virtue of its ancestral branchial derivatives: the second visceral arch giving rise to the inferior horn of the hyoid, the third to the greater horn, fourth and fifth to thyroid cartilage and sixth and seventh possibly to the arytenoid and cricoid cartilages (embryological development is discussed in section 4.1). The epiglottis, present in all mammals and serving respiratory, olfactory and deglutitive functions, evolved from tissue independent of the branchial arches. By establishing a connection between laryngeal entrance and nasopharynx it created a connection between upper and lower airways, which in most non-mammalian vertebrates are separated by the mouth or pharynx (Wind, 1976).

The vocal abilities and respiratory requirements of mammals vary between and within individual species. The laryngeal framework, however, is consistent comprising thyroid, two arytenoid and cricoid hyaline cartilages articulated at synovial joints and suspended by muscles and ligaments from the mandible and to a variable degree, skull base. Corniculate, cuneiform and epiglottic cartilages unlike the main framework are composed of flexible elastic cartilage, and according to Fink & Demarest (1978) form an integral part of the laryngeal system of 'springs'. Division of the hyoid complex into hyoid and thyroid in more primitive marsupials and monotremes is to some extent negated by fusion of thyroid to cricoid, yet the mobility between these two cartilages found in all other mammals is possibly the most important factor in production and modulation of voice. Evolution of the thyroid cartilage in mammals resulted in the development of variably controlled thyroarytenoid folds that allowed both laryngeal and respiratory development. Despite considerable differences in size, vocal ability and life-style, laryngeal morphology remains remarkably constant within the Mammalia. Variations in laryngeal framework are often subtle and inexplicable and reflect perhaps, deviations within species rather than major structural changes. Quantification of such variations is rarely possible, however, because of the scarcity of knowledge even of humans. Differences described in this section therefore consist of data derived from relatively small numbers of individual species and may not reflect major interspecial modifications.

Thyroid cartilage

In most respects the human thyroid cartilage provides a model for the majority of mammals in that it is composed of two plates of hyaline cartilage, joined anteriorly to form a protective shield to the airway. The posterior margin of each lamina carries superior and inferior projections (or horns) attached to the hyoid bone above and suspending the cricoid cartilage below. Five elastic ligaments are attached to the posterior aspect of the thyroid angle; median thyroepiglottic ligament, bilateral vestibular folds and the paired vocal folds.

Whereas the basic anatomy of this cartilage (and the remainder of the laryngeal skeleton) is well documented there is little information regarding the range of normal distribution in shape, size and configuration of individual cartilages or the size, symmetry and placement of the articulations and associated ligaments. Maue & Dickson (1971) measured the normal distribution of these parameters in 10 adult male and female human larynges. All linear measurements were made with a Vernier calliper correct to 0.01 mm and repeated by both investigators to ensure good reproducibility.

Although they found variability between the anterior height of the thyroid cartilages between male and female (the male being always larger), correlation studies confirmed only two measurements as being of significance. One of $r^2 = 0.97$ between the total horn height on the sides of the female larynges and 0.88 between the separation of the superior horns at their tips in male larynges. The true significance of these conclusions was questioned due to the low number of degrees of freedom associated with these measurements. Examination of the cricothyroid facets showed an absence on one side in 20% of the larynges with bilateral asymmetry and poorly defined boundaries a constant feature for all the larynges. Posterior cricothyroid ligaments supporting this joint were however, a constant feature. In humans it is recognized that many have asymmetrical larynges but clinically it is important to distinguish normal variations from pathological disturbance. Hirano *et al.* (1989) measured 17 dimensions and four selected angles in the framework of 50 fresh Japanese cadaver larynges. Ten were from newborns, 20 from adults below 30 years of age and the remainder from individuals over 50 years. All groups had equal numbers of male and female thyroid cartilages. They concluded that all dimensions and angles differed to some extent between the right and left sides in most larynges. The degree of asymmetry, however, did not differ significantly within the age groups or sexes although there was a suggestion that with age the right thyroid lamina tended to tilt laterally whilst the left tilted medially. It was suggested that this right sided deviation might be associated with the right-handedness present in all adults studied.

Comparative studies
The shape of the thyroid cartilage appears to be fairly constant in all mammals although modest individual variations are common without obviously influencing the purpose of this important cartilage. Zrunek *et al.* (1988) compared the

dimensions of 22 sheep larynges (following removal of all soft tissue to allow reproducible points of reference) with 21 human specimens. Having compared seven measurements they concluded that the adult sheep larynx falls within the range between male and female human larynges. This applied to values for height, anteroposterior diameter and the inferior breadth of the thyroid cartilage. Apart from the large arytenoids common to most herbivores, the general configuration proved to be very similar to that of the human larynx, except for an absence of any sexual variation in size. The organ was found to be large in relation to body size, this being apparently based on measurements of the vertical height of the thyroid laminae.

Comparative studies were carried out by Ibrahim & Yousif (1991) on the larynges of 20 adult goats and sheep. Although they found no correlation between cartilage size and body length or age of the animals, they concluded that the goat larynx was bigger in every respect than sheep of equal weight.

Saber (1983) compared the sheep larynx with that of the camel (*Camelus dromedarius*) finding that, except for fusion of the corniculate cartilage processes dorsally in the dromedary (as it is in the rabbit and many pigs), both larynges were typical of other domestic ungulates. Solis (1976) examined the laryngeal skeleton of the Philippine water buffalo (*Bubalus mindorensis*) concluding that although it resembled other cattle in shape, situation and attachments, it was bulkier, broader and shorter than that of other cattle and individual cartilages were thicker. Variations in weight and age were not considered nor were specific measurements included in this paper. Although a small number of papers have been published on the morphology of the mammalian larynx these are primarily descriptive and invariably confined to a single species. As part of the examination of the structure of the cricothyroid muscle, measurements were made of the area of the thyroid lamina in 47 different species of mammals. These ranged in size from the rock hyrax (*Heterohyrax brucei*) weighing 1.33 kg with a thyroid lamina area of 2.4 cm² to the hippopotamus of 3800 kg with an area of 64.1 cm². Although it was not expected that a significant relationship would exist between the area of a thyroid lamina and body weight, regression analysis was carried out for those animals with a body weight of under 100 kg. Analysis of the total group had shown wide scattering for animals above this weight although the 39 animals analysed had a body weight of between 1.33 kg and 96 kg, whilst the thyroid lamina varied between 2.4 cm² and 82.75 cm² in area (Appendix 2). Regression analysis at the 95% confidence level gave a correlation coefficient of $r^2 = 0.583$ ($r = 0.340$) suggesting the absence of significant positive allometry with regard to body weight and the area occupied by the thyroid laminae.

All specimens serially sectioned were scanned for morphological anomalies in the thyroid lamina. Differences in the frequency and pattern of ossification were common and are discussed in detail later in this section. The general shapes of the laminae were remarkably constant in larynges large enough to be

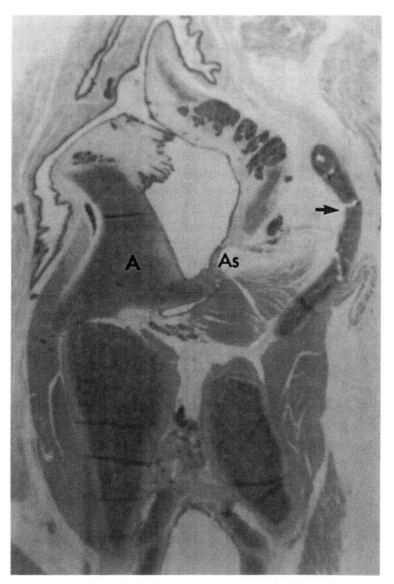

Figure 4.3. Sagittal section of the larynx of an adult agouti (*Dasyprocta leporina*) showing anterior bowing of the thyroid cartilage (arrowed), arytenoid (A) and anterior air sac (As).

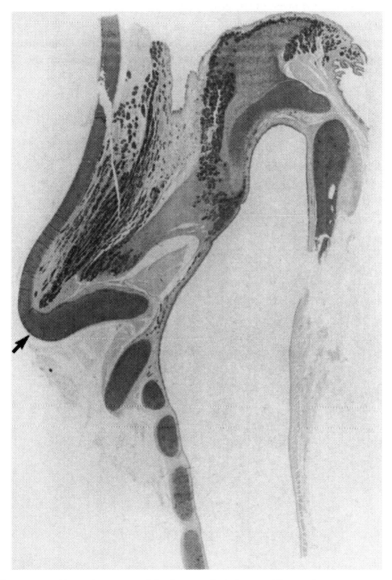

Figure 4.4. Sagittal section of the larynx of adult oryx (*Oryx tao*) showing marked anterior bowing of thyroid cartilage (arrowed).

visualized in three-dimensions. Several species showed marked anterior 'bowing', such as oryx (*Oryx tao*), agouti (*Dasyprocta leporina*), wallaby (*Macropus rufogriseus*), water chevrotain (*Hyemoschus aquaticus*) and tree shrew (*Tupaia tana*) (Figures 4.3 and 4.4). Although this modification increases the glottic area its value is uncertain, for none of the animals is particularly active or vocal. Indeed, the wallaby is largely aphonic.

Fenestra were found in either one or both laminae in a small number of specimens. Passing through the whole thickness of the cartilage the fenestra

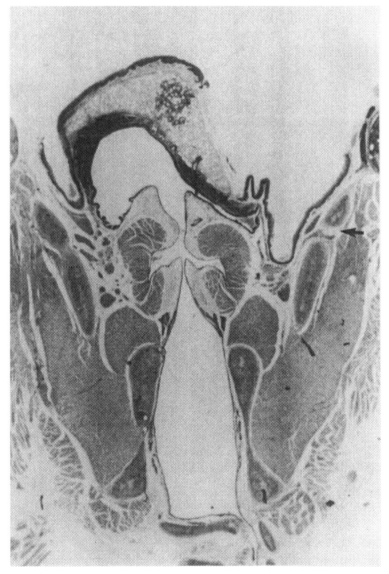

Figure 4.5. Coronal section to show fenestra in the left thyroid cartilage (arrowed) of adult mouse lemur (*Microcebus murinus*).

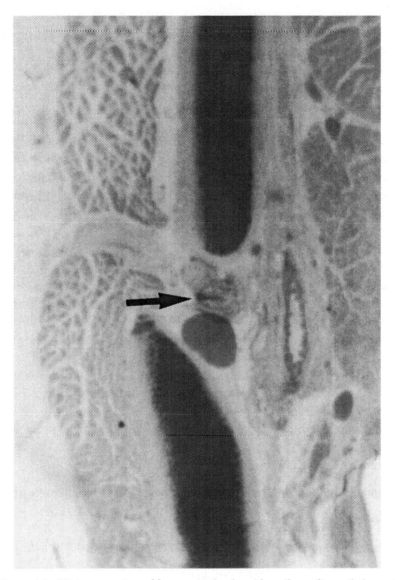

Figure 4.6. High power view of fenestra in the thyroid cartilage of juvenile leopard (*Panthers pardus*). The defect appears to be filled with a vascular bundle (arrowed).

contained blood vessels in the mouse lemur (*Microcebus murinus*), striped possum (*Dactylopsila trivirgata*), leopard (*Panthers pardus*) and porcupine (*Coendou prehensilis*) (Figures 4.5 and 4.6). In two specimens of the nine-banded armadillo (*Dasypus novemcinctus*) bilateral fenestra were filled with fibres from extrinsic muscle. These openings appear to be of no functional significance and may be defects of development. Examination of larger numbers of sections would probably reveal occasional fenestra in the upper part of the thyroid laminae of all mammals and have been recorded in adult human larynges by Harrison (1983). It has been called the foramen throideum and thought to represent the space between the fourth and fifth arches.

Significance of external laryngeal dimensions   Most mammals possess the capability of producing sound, some of considerable complexity. The pitch primarily relates to vocal cord length and the fundamental frequency and harmonics are then amplified to a variable degree by resonating chambers within the upper airway and possibly the nasal sinuses when present. Williams & Eccles (1990) suggested that the average fundamental frequency of the larynx could be predicted from external measurements of the laryngeal size. In 21 human cadavers they correlated the length of the membranous vocal cord with the distance between the lower border of the cricoid cartilage and thyroid notch. A correlation coefficient of 0.94 (p <0.001) was obtained and similar estimations were then made in 115 healthy volunteers between the ages of 31 and 71 years. Significant differences were found between males and females with respect to laryngeal size and vocal fundamental frequency. A correlation of 0.8 (p <0.001) was found between the external laryngeal size and speaking fundamental frequency for both sexes. The mean external size of the larynx was 46.1 mm $\pm$ 0.46 (range 36–55.5 mm) in males and 38$\pm$ 0.34 mm (range 31–42 mm) in females. Mean length of the vocal cord was 13.39 $\pm$ 0.52 mm (range 11–17.5 mm) in males and 9.61 mm (range 8–11.5) in females. From this data they derived an equation that for males, the:

Fundamental frequency (Hertz) = 190 − (external laryngeal size in mm) × 1.7

For females, the:

Fundamental frequency (Hertz) = 260 − (external laryngeal size in mm) × 1.4

Factors such as vocal fold mass and elasticity also influence laryngeal fundamental frequency and a repetition of the measurement of laryngeal size and vocal cord length on 52 sagittally sectioned non-humanoid larynges, produced a correlation coefficient of only $r^2 = 0.49$ (r = 0.31). With a large number of different species whose vocal cord length varied from 8 mm in the grey kangaroo (*Macropus* sp.; laryngeal size 20 mm) to 51 mm in a jaguar (*Panthera onca*; laryngeal size 82 mm) it was unlikely that a significant correlation would be obtained. Vocal cord bulk varies considerably within individual mammals as does voice production and this will be considered later in the section on laryngeal physiology.

The relationship of laryngeal dimensions to body size and gestational age in small infants is of importance when selecting the correct size of endotracheal tube. Schild (1984) correlated a variety of laryngeal measurements with crown–heel length in 27 infants at autopsy. Gestational age ranged from from 24 to 40 weeks. Computer generated curves from this data suggested that external measurement from thyroid notch to the lower border of the cricoid related well to crown–heel and crown–rump length and reasonably well to glottic length. Positive allometry is not present with regard to body and laryngeal size in this age group for as body size increases the larynx gets larger but at a slower rate. Glottic size is only one of the measurements necessary for correct choice of an endotracheal tube and there is some evidence that in this age group cricoid diameter is smaller than glottic area.

Cricoid cartilage

In the majority of mammals the shape of the cricoid cartilage is similar to that of the human. Roughly circular in shape inferiorly, and above, it follows the outline of the glottis. A broad posterior plate serves as site of origin for the posterior cricoarytenoideus muscle, in larger animals this frequently has a midline crest to increase muscle area. The anterior part of the arch is narrow and is joined above to the thyroid cartilage by the cricothyroid membrane. Inferiorly, at the junction with the trachea which is of similar circumference, the attachment is by a dense membrane attached to the lower border of the cricoid. Muscle attachments are described elsewhere.

Maue & Dickson (1971) analysed the anatomical variations in the laryngeal cartilages of 20 specimens and found few significant correlations in the cricoid cartilage measurements. There were differences of less than 1 cm in the anteroposterior and lateral dimensions. As in the thyroid cartilage, the male larynges exceeded the females in all dimensions and cricothyroid facets were so variable in size, shape and presence that accurate quantification was impossible. Thirty per cent showed only soft tissue facets and when removed no obvious cartilaginous or bony recesses were revealed. Only 20% had well defined facets; these were always bilateral and associated with facets on the corresponding thyroid horn. The detailed morphology of the cricoarytenoid joints will be considered separately in this section.

Comparative studies

The morphology of the diminutive cricoid cartilages found in the Chiroptera and other small mammals can be studied by constructing cardboard models from tracings of serial sections. Figure 4.7 shows the cricoid cartilage of the greater horseshoe bat (*Rhinolophus ferrumequinum*) with its large sagittal crest, a feature common to most bats. As with other laryngeal cartilages in Chiroptera, the cricoid is ossified early in life. Variations in shape are common within individual species and are thought by Denny (1976) to be related to the presence or absence of air sacs (Figure 4.8).

In other non-human mammals cricothyroid joints vary in size and may be absent: in marsupials no joints were found in serially sectioned specimens. Fusion of the anterior cricoid arch to thyroid cartilage is said to occur in all marsupials and monotremes although this appears to be a combination of thickening of the cricothyroid membrane and direct cartilaginous union (Figure 4.9). A much larger number and variety of species from these orders need to be studied before an accurate three-dimensional assessment can be made to confirm the morphology.

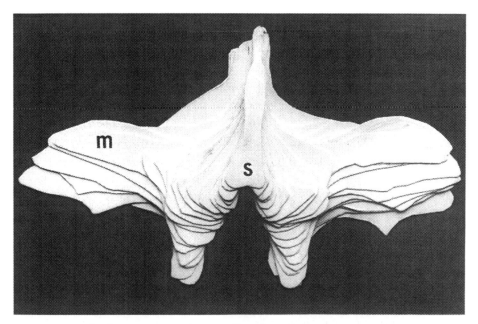

Figure 4.7. Reconstruction from cardboard sections of serial drawings of the cricoid cartilage of greater horseshoe bat (*Rhinolophus ferrumequinum*): muscular 'wings' (m) and posterior spine (s).

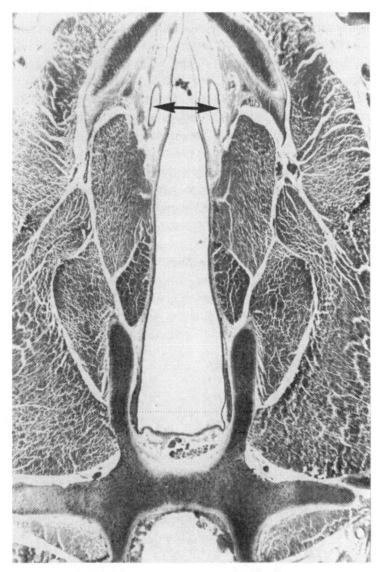

Figure 4.8. Transverse section of the larynx of the moustached bat (*Pternotus parnelli*) to show large cricoid cartilage and intrinsic musculature. (Ventricles arrowed.)

As an integral part of an enquiry into the influence of laryngeal and tracheal size on the running speed of mammals, the maximum area between the vocal cords and also within the first tracheal ring were measured in 129 animals (Harrison & Denny, 1985). (Since the area within the cricoid cartilage is almost identical to that of the first tracheal ring, care must be taken with species where the trachea is capable of expansion on inspiration.) This provided comparative data for a variety of species of varying ages and weights and is given in Appendix 3, with the area between the vocal cords measured at maximal abduction, using tension sutures. The significance of these measurements in relation to maximum respiratory effort is discussed in the section on locomotion but in each instance the glottic area was less than that of the cricoid cartilage in the adult animal. The data shown in Appendix 3 does provide a guide to the sizes of endotracheal tube that were found to be suitable for individual species, the ranges representing variations in weight and maturity.

It is well known that the largest animals are not the fastest, indeed maximum running speed (MRS) is highly variable even among mammals of similar size and within the same family. Table 4.2 shows least squares linear regression analysis of $\log_{10}$ transformed weight versus tracheal (cricoid) area for 129 species and groups from individual families. Significant relationships appear to exist for the 20 deer ($r^2 = 0.790$) and 11 zebra ($r^2 = 0.664$) suggesting that in these animals the cricoid area increased in size as the animal became heavier. However, the potential errors in such an analysis are well recognized with reference to both correlation coefficients and regression analysis. The former

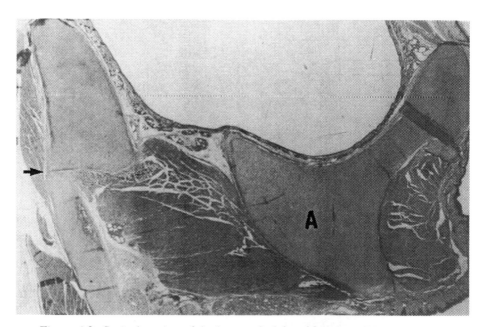

Figure 4.9. Sagittal section of the larynx of adult red kangaroo (*Macropus rufus*) showing large arytenoid (A) and ?fusion of cricoid and arytenoid (arrowed).

Table 4.2. *Least squares linear regression analyses of log_{10} transformed weight (kg) versus tracheal area (cm^2)*

Group	n	Intercept (a)	Slope (b)	Significance	Correlation r^2	Significance
All mammals	129	−0.200	0.4180 ± 0.035	0	0.535	7.001
Antelopes	26	0.582	0.4935 ± 0.072	0	0.664	7.001
Deer	20	0.593	0.5380 ± 0.065	0	0.790	7.001
Zebra	11	1.841	0.0324 ± 0.138	0.8135	0.664	7.02
Pigs, etc.	13	0.541	0.1675 ± 0.090	0.088	0.238	0
Camels, etc.	10	0.365	0.2811 ± 0.088	0.0126	0.559	7.01
Cats	11	1.436	0.7434 ± 0.176	2.6E−03	0.664	7.001
Bears	11	3.197	0.1214 ± 0.360	0.7419	0.0125	0
Primates	7	0.1065	0.2477 ÷ 0.097	0.0501	0.5668	7.02
Wolves	7	0.0129	0.4344 ± 0.186	0.0653	0.5229	7.05

measures the degree of linear association between two continuous variables, but may be artificially high if a few observations are very different from the rest. This has been minimized by examining scatter plots prior to analysis. Regression is used to estimate a dependent relationship between one variable and another, the choice of which is the 'dependent' and which the 'independent' variable can markedly effect the result. The standard error of the slope and intercept where appropriate, is therefore important in presenting results.

Arytenoid, epiglottis and associated cartilages

When examining the human arytenoid cartilage Maue & Dickson (1971) found noticeable symmetry between right and left side and in the distances between apex to vocal and muscular processes. Although ratios remained constant between the sides, sex differences were prominent as in other laryngeal cartilages. As with the thyroid and cricoid, the arytenoid (except possibly for the body) is composed of hyaline cartilage, whilst the body, epiglottis and corniculate and cuneiform cartilages are elastic. Fibro-elastic cartilage differs from hyaline cartilage mechanically in that it resists tension whereas hyaline cartilage resists compression.

The human epiglottis is described as being a centrally placed, broad flat plate with a superior margin and free lateral edges. At its base it narrows with a leaf-like shape and is attached to the top of the thyroid cartilage by a ligament. The body is perforated by numerous pits filled with glandular tissue providing an entry into the associated pre-epiglottic region.

All mammals possess an epiglottis although it varies considerably in size, presence of pits and glandular tissue. Apart from the higher apes where the epiglottis is relatively small and not intranarial, it probably serves for both olfaction and supraglottic closure. Its size, shape and glandular content appear to be unrelated to the life-style of individual species. For example, the northern night monkey (*Aotus trivirgatus*) has an extremely large epiglottis with many pits and glands but does not have a keen olfactory sense. When a good sense of smell is required, however, as in the Carnivora, the epiglottis is always large. During swallowing the epiglottis is thought to be pulled up towards the hyoid and against the tongue base. In some animals this may be assisted by the presence of a linguo-epiglottic muscle (Figure 4.10). Examination of coronal serial sections showed this muscle in most carnivores as well as in the wallaby (*Macropus rufogriseus*, *M. agilis* and *M. parma*), hairy and nine-banded armadillo (*Chaetophractus villosus* and *Dasypus novemcinctus*), saki monkey (*Pithecia pithecia*), tree shrew (*Tupia tana*), tree and rock hyrax (*Dendrohyrax arboreus* and

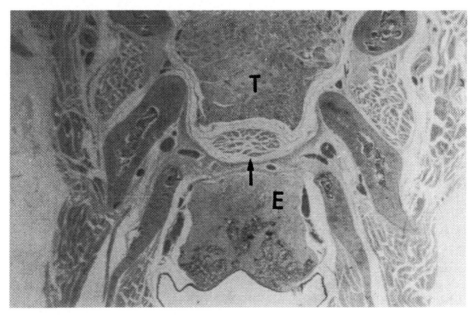

Figure 4.10. The linguo-epiglottic muscle (arrowed) of the nine-banded armadillo (*Daspus novemcinctus*). Epiglottis (E) and tongue base (T).

Heterohyrax brucei) and crested porcupine and acuchis (*Hystrix cristata* and *Myoprocta pratti*).

The role of the arytenoid and associated cartilages in the functional anatomy of the immature larynx

The morphology of the larynx appears to be determined in part by the food that each animal takes, although the basic design remains essentially similar. An intranarial epiglottis whilst enhancing olfaction permits respiration during feeding, which is particularly desirable in ruminants. In all mammals the vocal cords stretch from thyroid cartilage to arytenoid, on the tip of which lies the corniculate cartilage (of Santorini). The cuneiform cartilage (of Wrisberg) when present is related to the anterior aspect of the arytenoid and base of the epiglottis. Although a constant feature in every mammalian larynx the size and shape of each of these structures depends upon the age and species of the animal. A large arytenoid surmounted by a large corniculate cartilage provides a high posterior wall to the larynx. Steep lateral walls are achieved by mucosal folds between epiglottis and arytenoid supported by cuneiform cartilages. Both allow masticated food to flow past the centrally placed larynx avoiding overspill.

Table 4.3. *Table of measurements for 13 human immature larynges*

Age	1	2	3	4	5	6	7	8
1 day	2.8	3.8	7.7	5.4	5.0	14.0	11.0	–
8 days	2.5	0.0	9.3	4.8	3.9	10.0	9.0	–
6/12	4.8	2.5	13.6	6.0	4.6	12.2	12.2	–
11/12	3.9	2.5	12.1	5.0	4.2	11.0	9.3	–
6/52	3.6	2.7	13.5	6.1	4.6	13.5	11.5	3.0
5/52	4.4	1.8	10.0	5.2	4.0	9.0	6.9	1.3
7/52	4.3	2.1	9.0	4.0	3.8	11.5	9.5	1.8
2/52	3.7	0.5	9.3	6.3	3.6	13.0	11.0	3.7
2/52	3.7	3.2	12.3	3.9	3.6	13.0	12.8	4.3
1 day	3.7	1.8	10.2	4.5	4.5	14.0	11.6	2.8
6/52	5.3	1.9	9.2	6.5	4.0	12.0	12.0	2.6
3 days	6.1	3.1	7.91	4.7	4.5	9.7	9.8	2.9
3 days	4.1	1.8	8.3	5.5	3.7	11.2	8.4	2.8

1. Distance between the curled extremities of epiglottis.
2. Distance between cuneiform tubercles.
3. Distance from epiglottic tip to upper border interarytenoideus.
4. Diameter of trachea in coronal section.
5. Diameter of trachea in sagittal section.
6. Distance from epiglottic tip to anterior larynx (transverse plane).
7. Distance from Anterior larynx to cuneiform swelling.
8. Distance from Top of corniculate cartilage to upper border interarytenoideus.
All measurements in mm.

Such modifications are unnecessary in carnivores where food is swallowed whole without chewing. Clearly the manner in which the laryngeal inlet is protected depends to a considerable extent on the nature of the food ingested and the broad principles on which this is based are discussed in Chapter 3. Pracy (1984) studied these structural variations in detail in both the immature and mature mammal, using specimens from those listed in Appendix 1.

The morphology of the infant larynx is entirely different from that of the adult. This is due to the relative difference in size of the epiglottis and arytenoid cartilages and also to distension of the submucosal glands. In the human infant the cuneiform swelling is relatively more prominent than it is in the adult. However, the cuneiform cartilage is not present in the newborn; it can be identified only after six months when it is found to be 'kite shaped' with its posterior edge applied to the anterior border of the arytenoid cartilage. It appears that in the infant this cartilage plays an important role in supporting the aryepiglottic fold, thus maintaining a high lateral wall for the supraglottis. The epiglottis is usually curved with the two margins of the petiole approaching the midline posterior to the body of the cartilage. This causes the aryepiglottic folds to lie close together and, with the increase in associated glandular tissue, provides extra protection for the glottis.

Thirteen infant larynges were measured and this data is shown in Table 4.3. The relative differences in height of the arytenoid and cricoid cartilages were calculated by comparing the height of an arytenoid with that of the cricoid in sagittal section. All young mammals are fed on milk alone for a certain period; weaning times vary depending on the life-style of the species and future diet. Many herbivores suckle for only a short time before adding vegetatation to their diet, although data based on captive animals may be inaccurate due to prolonged weaning and lack of environmental danger.

Measurements were carried out by Pracy (1984) to test the hypothesis that the posterior laryngeal wall was higher in adult herbivores and omnivores than in carnivores, because of a need for the former to eat a largely liquid diet. Accurate evaluation of the height of the arytenoid component in the carnivore could be obtained by measurement from the upper border of the cricoid to the arytenoid tip. This is not possible in Artiodactyla and Perissodactyla where the vocal process of the arytenoid lies below the upper border of the cricoid. Measurements were therefore made from the lower surface of the vocal fold to the tip of the arytenoid in these specimens. Data from 29 adult animals is collated in Table 4.4 confirming that the height of the posterior laryngeal wall is lower in carnivores than herbivores. Table 4.5 includes measurements for 19 juvenile or baby larynges showing that in the young 'milk drinking' animal the posterior laryngeal wall tends to be high irrespective of their future eating habits.

Males bigger, females biggest

In most species one sex is bigger than the other and this is generally the male (e.g. sheep, cattle, pigs, weasels, deer, etc.; Reiss, 1982). Often the degree of dimorphism is slight, although the average elephant sea bull is eight times heavier than the cow. One explanation for this variation is the need for the male to compete for access to females for breeding. Yet the male Weddell seal (*Leptonychotes weddelli*) may hold a harem of up to 12 females although the female is significantly the larger animal. However, the Weddell seal must nourish her pup in hazardous conditions and, therefore, a larger female has more reserves to

Table 4.4. *Adult animals measured for calculating arytenoid/cricoid ratios*

Species	Order	Arytenoid	Cricoid	Ratio
Asiatic black bear	Carnivora	1.84	2.81	0.65
Cheetah	Carnivora	0.63	2.68	0.24
Leopard	Carnivora	1.65	3.16	0.52
Puma	Carnivora	1.12	2.65	0.42
Wolf	Carnivora	1.5	2.79	0.53
Sloth bear	Carnivora	1.6	2.95	0.54
Red panda	Carnivora	0.35	0.9	0.38
Coati	Carnivora	0.45	1.15	0.39
Cheetah	Carnivora	5.4	11.1	0.48
Gorilla	Primate	2.82	2.95	0.95
Chimpanzee	Primate	1.35	1.5	0.90
Gelada baboon	Primate	5.7	9.7	0.59
Sooty mangabey	Primate	6.8	7.0	0.97
?Vervet monkey	Primate	7.6	6.2	1.22
Zebra	Perissodactyla	3.69	3.9	0.94
Tapir	Perissodactyla	4.74	4.87	0.97
Onager	Perissodactyla	4.99	5.25	0.95
Mountain zebra	Perissodactyla	4.08	4.37	0.93
Waterbuck	Artiodactyla	3.09	4.81	0.64
Pere David deer	Artiodactyla	4.32	4.64	0.93
Gemsbok	Artiodactyla	3.8	4.9	0.77
Dromedary camel	Artiodactyla	5.17	5.69	0.9
Wild boar	Artiodactyla	3.88	3.9	0.99
Timor deer	Artiodactyla	2.28	3.36	0.67
Collared peccary	Artiodactyla	2.26	2.48	0.91
Muntjac	Artiodactyla	1.65	1.65	1.0
Fallow deer	Artiodactyla	3.76	4.47	0.84
Markhor	Artiodactyla	2.87	3.17	0.9
Yak	Artiodactyla	3.0	5.72	0.52

achieve this. Another factor may lie in quantitative analysis of nutritional factors and energy needs. This increases almost linearly with body weight and in females is maximized for reproduction; while some males simply have energy available for reproduction! In monogamous species males typically invest heavily

Table 4.5. *Babies and juveniles measured for arytenoid/cricoid ratios*

Species	Order	Arytenoid	Cricoid	Ratio
Lion	Carnivora	2.72	4.71	0.57
American black bear	Carnivora	1.49	3.02	0.49
Tiger	Carnivora	1.63	2.77	0.58
Brown bear	Carnivora	1.29	2.04	0.63
Badger	Carnivora	4.6	6.19	0.74
Serval	Carnivora	2.8	7.15	0.39
Lar gibbon	Primate	1.9	1.8	1.05
Ruffed lemur	Primate	6.6	7.1	0.93
Zebra	Perissodactyla	2.9	2.13	1.30
Wild horse	Perissodactyla	2.6	2.19	1.18
Kudu	Artiodactyla	3.21	4.34	0.73
Musk ox	Artiodactyla	3.73	4.14	0.9
Nilgai	Artiodactyla	3.6	3.42	1.05
Blesbok	Artiodactyla	3.14	3.31	0.94
Pygmy hippopotamus	Artiodactyla	7.7	8.7	0.88
Oryx	Artiodactyla	1.87	1.65	1.13
Mouflon	Artiodactyla	9.0	12.4	0.73
Oryx	Artiodactyla	8.2	8.8	0.93
Sealion	Pinnipedia	2.78	2.78	1.00

in their offspring and this results in males and females being about the same size.

Whatever the underlying explanation, in most species the adult male larynx is quantitatively larger than the female, with a corresponding difference in sound production. This disparity is characterized by a general increase in all dimensions within the framework of the male larynx and is assumed to be a manifestation of androgen stimulation. Relationships between the presence of testes and development of secondary sexual characteristics such as voice change and prominence of the thyroid cartilage are well recognized. The introduction of the hormone-receptor concept has led to the confirmation of high affinity androgen receptors in all the laryngeal cartilages (Tuohimaa *et al.*, 1981). Androgen induced changes in laryngeal growth have been studied in young rams using varying doses of both testosterone and dehydrotestosterone (Beckford *et al.*, 1985). They found a positive dose-response relationship for the anteroposterior diameter, superior thyroid horn height, posterior thyroid cartilage width, thyroid cartilage angle and vocal process to arytenoid base distance. The thyroid cartilage demonstrated the greatest response to androgen stimulation although this may have been related to the duration of androgen exposure. The thyroid angle is generally recognized as the classic example of sexual dimorphism in the human. A prepubescent angle in both male and female of 88° changes postpuberty to over 120° in the male, and 100° in the female.

However, such dimorphism is not common to all mammals and is not solely the response to androgen stimulation. Audemorte *et al.* (1983) and Holt *et al.* (1986) studied oestrogen receptors in the larynx of the baboon (*Papio cynocephalus*) using an autoradiographic technique with tritiated oestrogen. Although no receptor activity was found in surface epithelium, all tissues of mesenchymal origin had consistently high levels of nuclear localization of labelled oestrogen. This was highest at the anterior commissure and in the laryngeal cartilages and perichondrium. Sex steroid receptors are proteins which bind with specific hormones in the cell's fluid compartment. They form complexes that, following concentration within the nucleus, interact with the genetic material to influence protein synthesis. The strongly labelled nuclear uptake in laryngeal (and tracheal) cartilages and perichondrium suggests that changing hormone levels could influence the competency of laryngeal muscles, ligaments and cartilages. This may be of particular importance in the ageing animal but could also be an oversimplification of a complex and species specific situation.

It has long been known that the male and female spotted hyaena (*Crocuta crocuta*) have similar levels of circulating androgen even in the foetal stage, thus offering an explanation for the dominance exerted by the larger female over the male. It is now known that this balance persists only until puberty, when the male develops about six times as much circulating hormones as the female. However, this level is dependent upon 'rank' and the highest ranking female may have more androgen than the average male. Growth of the mammalian larynx therefore appears to be controlled by both androgen and oestrogen, possibly by modulation of genetic expression. This does not offer a rational explanation for the variations in sexual dimorphism which are a feature of this order.

Structure and ossification in laryngeal cartilages

The appearance of the cartilaginous plate in the evolutionary history of the mammalian larynx heralded the importance of collagen as a factor in establishing the strength of this organ and the integrety of its architecture. Cohen *et al.*

(1993) examined, by biochemical techniques, the collagen composition of the normal developing human larynx and trachea in 28 humans, ranging in age from birth to 44 years.

The three interstitial types of collagen (I, II, and III) are found in connective tissue with type II predominant in hyaline cartilage. Histologically, all the laryngeal cartilages with the exception of the epiglottis and the medial edge and tip of the arytenoid are composed of hyaline cartilage. They found however, that type II collagen was only found in large amounts in the arytenoid, hyoid and cricoid cartilages. This amount gradually decreased after the age of 12 years but small amounts of synthesized collagen were still found even in mature larynges, indicating significant turnover and remodelling.

Ossification

Osseous changes within the human larynx are often considered to be a degenerative process in hyaline cartilage associated with advancing age. Woven bone is formed first, followed by lamellar bone following the normal process of endochondral ossification as found in the long bones. Rigid separation between calcification and ossification is impossible, for when cartilage is transformed into bone, calcification precedes ossification. Areas of calcification then disappear as ossification spreads to these regions. Examination of 145 serially sectioned larynges removed for carcinoma in patients of both sexes between the ages of 19 and 74 years, has confirmed these changes (Figure 4.11). Although ossification was seldom evident until the third decade, absorption and vascularization of cartilage can be seen in the second decade of life, particularly in the posterior part of the thyroid lamina. Calcification, unaccompanied by ossification, may now occur with calcium deposited around the capsule of the cartilage cells. Such calcified foci or plaques are the precursors of ossification and invariably disappear with the spread of bony changes. The presence of bone marrow can be seen in the early stages of ossification and occasionally bone marrow tissue is found in cartilage without bone formation.

The pattern of ossification in the human larynx has been studied radio-logically in 516 patients by Hately et al. (1965) and by a combination of radiology and tissue sectioning in 133 Black African larynges by Keen & Wainwright (1958). Their conclusions are similar to those found by Harrison & Denny (1983) in their radiological examination of 341 larynges, 145 of which were serially sectioned and 86 macroscopically cut laryngectomy specimens, together with 110 normal larynges. Ossification changes usually occurred in the third decade but in only one in this series, a girl aged 18 years, was it found before the age of 21. Rare instances of laryngeal calcification in childhood have been recorded apart from cases of chondrodystrophia calcificans congenita. Although it is possible to demonstrate a gradually increasing degree of ossification for both sexes with advancing age, direct correlation between age and degree of ossification is poor. Instances of almost total ossification can be seen in individuals no more than 30 years of age, whilst in this group two men aged between 59 and 61 showed no ossification on both radiology and serial sectioning.

Due to a disparity between the number of male and female larynges examined it was impossible to be certain whether a sexual orientated difference existed when data was matched for age.

Patterns of ossification   Within the thyroid cartilage ossification begins, invariably, at the posterior border near the root of the inferior horn. Spreading along the inferior border it reaches the midline where it may unite with a separate centre of ossification. The process then extends upwards to the upper and posterior part of the thyroid lamina with the superior horn late to ossify. The superior margin remained unossified in 20% of sections and 'windows' of unossified cartilage may remain in the centre of the alar giving the appearance of neoplastic destruction (Figure 4.12).

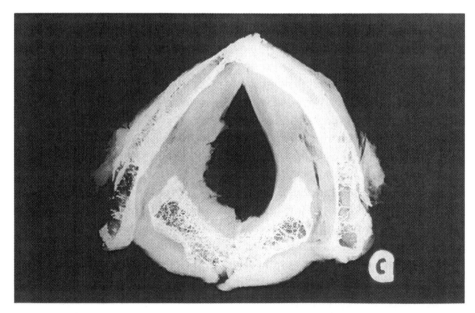

Figure 4.11. X-ray transverse section of a human larynx (aged 57 years) showing extensive ossification.

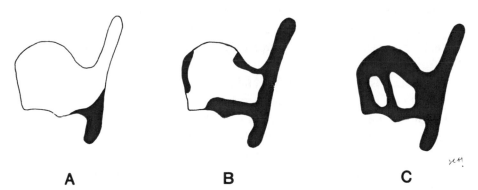

Figure 4.12. Schematic representation of patterns of ossification in human adult larynx – progressing from A to C.

Cricoid ossification occurs about the same time as in the thyroid, beginning at the upper border of the posterior plate. It spreads along the superior margin and is followed by an extension to the remainder of the ring with the anterior part of the arch last to ossify. Ossification of the arytenoids is more difficult to grade radiologically but is well shown in serial sections. Starting in the muscular process it spreads to the base but is seen rarely at the apex or vocal process. Although calcification has been described in the elastic cartilage of the pinna, giving rise to the term 'petrified auricle', ossification of the epiglottis is rare.

Heterotopic ossification is seen in the respiratory tract in the condition tracheopathia osteochondroplastica and the patient with epiglottic ossification reported by Milroy (1992) is possibly another example of heterotopic ossification. However, patches of ossification were seen in the epiglottis of an adult hog badger (*Arctonyx collaris*) (Figure 4.13).

Why does ossification occur?   Various propositions have been advanced to explain the stimulus which results in laryngeal ossification. Vastine & Vastine (1952) described the pattern of ossification in the larynges of five pairs of identical twins. Apart from one pair, aged 13 years, who showed no ossification, the remainder had identical patterns of calcification. Genetic control may influence the timing of ossification and rate of spread but correlation between increasing age and degree of ossification is poor. There is a possibility that the stimulus is provided by the mechanical effect of muscle contraction. Denny (1976) describing the larynges of Chiroptera found ossification of all cartilages to be common in Rhinolopoidea (horseshoe bats) and the Emballonuridae (sheath-tailed bats) at an early age, whilst the Phyllostomatoidae (New World leaf-nosed bats) showed lesser degrees of ossification. All belong to the suborder Microchiroptera and echolocate. By comparison with other mammals they have large intrinsic muscles, possibly requiring a rigid framework to function effectively (Figure 4.14).

The relationship between osteogenesis and tensile forces is clearly seen in avian tendons and in the tail of kangaroos. Ossification is influenced by tension and pressure although, under natural conditions, bones may develop complicated or even bizarre shapes without being subject to any obvious mechanical stimuli. It appears therefore that the appearance of centres of ossification depends entirely on intrinsic growth factors and it is possible to speak of the morphological 'individuality' of each bone or cartilage.

With this concept in mind Harrison & Denny (1983) have studied the pattern of ossification in non-hominoid primate larynges by combination of radiology and serial section. Representatives of all families have been studied although individual numbers are small. Seven specimens came from animals dying either at birth or before maturity; none showed evidence of framework ossification apart from that in the hyoid bone. In all remaining adult primates the pattern and degree of ossification (Figure 4.15) were remarkably similar to that found

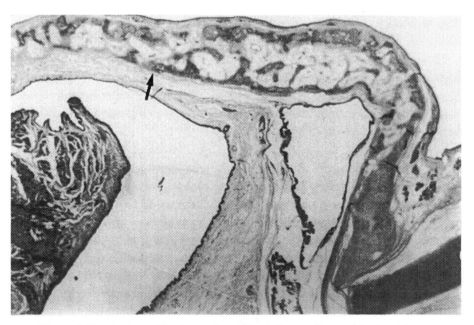

Figure 4.13. Patchy ossification in the epiglottis (arrowed) of hog badger (*Arctonyx collaris*).

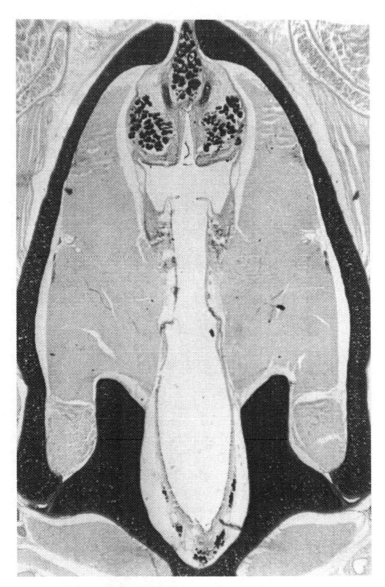

Figure 4.14. Ossification in all cartilages of the yellow-winged bat (*Lavia frons*).

in the human larynx. Most animals showed patchy ossification but realistic comparisons of 'natural' life expectancy time scales between different species (including humans) is difficult and many of these specimens may have come from relatively young animals. All larynges from the species listed in Appendix 1 were X-rayed, and the patchy ossification found in the cartilages of most of the adults was associated with sites of muscular attachment (Figure 4.16). The commonest sites for ossification and for neoplastic invasion are the areas of the laryngeal framework where collagen bundles from intrinsic muscles penetrate the perichondrium. These sites represent zones of increased tension leading

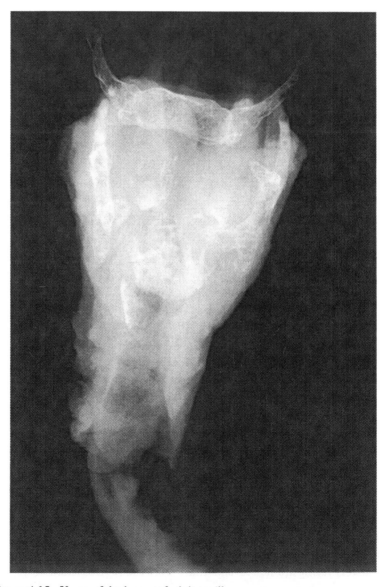

Figure 4.15. X-ray of the larynx of adult gorilla.

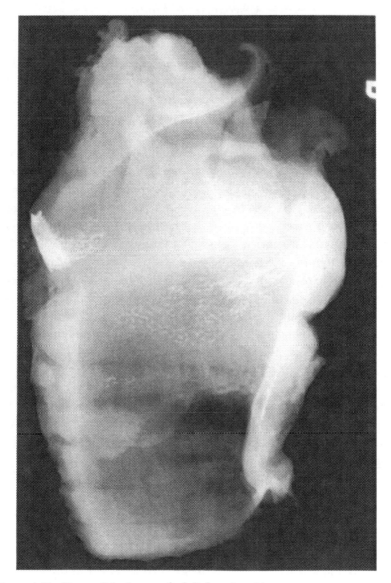

Figure 4.16. X-ray of the larynx of adult American bison.

eventually to replacement ossification. The speed with which this occurs might be influenced by factors such as age and physiological requirements, for some bats must produce audible sound, echolocate and respire whilst 'on the wing'.

Cricoarytenoid joint

The existence of a joint between the cricoid and arytenoid cartilage has been known since the time of Galen, yet details of its precise anatomy and function have remained controversial. Rotation around a vertical axis, lateral and medial sliding, rocking and synchronous three plane movements have all had their advocates. Final analysis has only come following fundamental research by Sellars & Keen (1978) and Sellars & Sellars (1983), who studied the precise anatomy of the human cricoarytenoid joint.

Each cricoarytenoid synovial joint is found between a facet on the inferior aspect of the muscular process of the arytenoid and an elongated facet on the upper aspect of the cricoid lamina. All 60 arytenoid articular facets were found by Sellars & Sellars (1983) to be concave and similar in shape and symmetry. The short axis was parallel to the long axis of the cricoid facet and shorter, in all observations, than the length of the cricoid facet with which it was articulated. The average difference between the lengths of these two axes was about 2 mm, the largest 2.5 mm (Sellars & Keen, 1978). At rest the long axis of the arytenoid facet was at right angles to the long axis of the cricoid facet. The latter was more or less oval in shape and each was convex on its short axis and convex on its long axis. Measurements of 60 cricoid facets found a mean medial curvature (radius of 2.1 mm) significantly smaller than the lateral curvature (radius 2.6 mm).

Asymmetry was evident between the cricoid facet axes of different larynges and often between the two facets in the same larynx. The positions of the two arytenoid vocal processes could therefore be asymmetrical. Despite this the ultimate movement of each arytenoid cartilage ensures correct and synchronous opening and closing of the glottis. Variations in individual joint morphology and muscle action suggest therefore, that movements are specific to each joint.

The joint capsule was found to be firmly attached to the rim of the articular facets of the arytenoid and cricoid cartilages. This fibro-elastic sleeve is lax and strengthened on its medial aspect by a strong oblique ligament attached below to the upper surface of the cricoid lamina medial to the cricoid facet. Above it reaches the medial surface of the arytenoid midway between the the vocal process anteriorly and muscular process posteriorly. Although more accurately called the 'cricothyroid ligament' it continues to be called the posterior cricoarytenoid ligament. Radiological studies (Ardran & Kemp, 1966) proved conclusively that a rocking action of the arytenoid cartilages occurred when they showed the inferiorly placed position of the vocal processes in adduction, and a higher position in adduction. Sellars & Keen (1978) therefore concluded that three movements are combined in the cricoarytenoid joint action; inward and outward rocking, medial and lateral sliding and rotation. Details of these movements and their role in vocal cord function are considered in Chapter 5.

Comparative studies   Following removal of the thyroid cartilage and surrounding muscles the cricoarytenoid joints were exposed in 22 formalin preserved specimens. After inspection of the joints and related ligaments the capsule was opened to expose the articular surfaces. At various stages in the dissections photographs were taken to record muscle attachments and drawings made of arytenoid and cricoid after measuring the length and midbreadth of the articular facets (Figures 4.17 and 4.18). Appendix 4 lists this data and shows that, unlike the human, in some species the long axis of the cricoid facet exceeded the opposing long axis of the cricoid by more than 2.5 mm. In the yak (*Bos mutus*), kudu (*Tragelaphus strepsiceros*) and Przewalski horse (*Equus przewalskii*) it is over 3 mm, in the lion (*Panthera leo*) more than 6 mm. With such a small number of specimens it is impossible to obtain more than a general impression of similarities and variations between different species. There is no reason for expecting that movement of the arytenoid on the cricoid is fundamentally different from that found in humans. In these dissections all arytenoid facets were situated on the inferior surface of the muscular process and approximately oval in shape. They were concave on the short axis but in most animals not particularly curved on the long axis, although this was variable. The cricoid facets were oval in shape but in some cases noticeably elongated and convex on both long and short axes (e.g., wolf and cheetah).

Minor differences in shape and curvature of the articular facets are to be expected but a larger more representative series would demonstrate their real significance. Despite variations in size and behaviour, however, dissections of the cricoarytenoid joints reveal a constant pattern in most mammals, reflecting similarities in laryngeal function.

The basic structure of the mammalian larynx clearly conforms to a fundamental pattern, with individual morphological modifications being related to specific functional needs. These are best illustrated by the larynges of the Cetacea and Chiroptera, who have succeeded in overcoming an aquatic and aerial environment without impairing the primary function of their larynges to protect the respiratory tract.

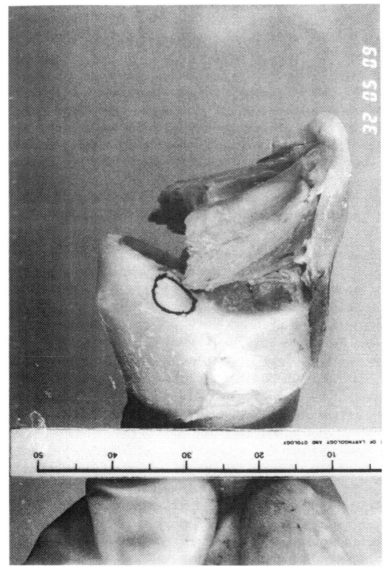

Figure 4.17. Cricoid cartilage of cheetah (*Acinonyx jubatus*) with arytenoid facet marked with gentian violet. (Rule in mm.)

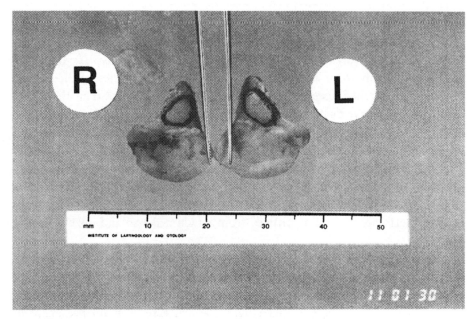

Figure 4.18. Arytenoids of cheetah (*Acinonyx jubatus*) with articular facets on the inferior surface of the muscular processes marked with gentian violet.

4.3 Ventricle, saccule and air sacs

Although many authors have credited Galen with the first description of the laryngeal ventricle there is no reliable evidence to support this claim. It is certain however, that he did recognize the cerebral ventricles. It seems certain that he worked largely with apes, dogs, horses and ruminants, transferring his results by analogy to the human body. Vesalius in both the 1543 and 1555 editions of *De Humani Corporis Fabrica* gives detailed accounts of the structure of canine and bovine larynges but makes no reference to the ventricles or for that matter, to the cuneiform cartilages or aryepiglottic folds (Garrison & Hast, 1993). Since his training was firmly rooted in the Galenic school of medicine it is unlikely that he would have omitted such structures accidentally. In 1600 Fabricius when considering the comparative anatomy of the laryngeal regions of sheep, ox, pig, horse and ape, correctly described the epiglottis, glottis and ventricles. Since his *Fabrica* concentrated primarily on human anatomy it seems likely that he also recognized the ventricle in human larynges (Fabricius, 1600). Despite this early work, credit for an accurate description of both ventricle and associated saccule is usually given to Morgagni. Born in Forli (Romagna) on February 25th, 1682, he held the Chair of Theoretical Medicine at Padua from 1709 to 1712 and in 1715 was appointed to the prestigious Chair of Anatomy in that University. His many anatomical works were published under the title of *Adversaria Anatomica* and in paragraph 16 of edition 1, he described the ventricles of the larynx. Following a detailed account of the glottis he proceeds 'between these two ligaments that I have made evident, there is a fissure of elliptical form present in one or the other side, and of such amplitude that we can partially introduce the tip of our thumb or other digit. Their lower portion is limited by the inferior fibres of the thyroarytenoid muscle . . . in that part where they become closer to the base of the epiglottis the cavities are deeper because of the addition of a larger appendix. These are covered by the same tissue which lines the larynx and are perforated by many orifices from which a dense and oozing humor extrudes'.

He also believed that Galen had recognized these cavities, calling them ventricles, but did not separate the ventricle from the saccule. Commenting that Fabricius and his pupil Casserius had described laryngeal ventricles in both horse and pig, Morgagni recognized that these structures were not present in all animals. He also criticized those comparative anatomists who extrapolate their findings in one species to all others, without first carrying out adequate dissection (Canalis, 1980). The first detailed description of the laryngeal saccule was given by Hilton in 1837, he called it the 'laryngeal pouch or sacculus'. His meticulous dissections of human larynges revealed this pouch extending upwards from the fossa elliptica (ventricle) for, on average, half an inch (13 mm). 'When distended it reached the upper edge of the thyroid cartilage and the opening below was protected by two small semilunar folds of membrane placed anteriorly and posteriorly with respect to the centre of the aperture. The pouch was perforated by numerous minute openings, the terminations to the excretory tubes from the glands which surround and belong to the the pouch' (Hilton, 1837). Slips of thyroepiglottic and aryepiglottic muscle fibres envelope the saccule and he assumed that these exerted compression during phonation, lubricating the vocal cords.

Hilton was apparently unaware of Morgagni's earlier account of the ventricle and possibly the saccule but his attention was drawn to this omission by the editor of the Guy's Hospital Reports. His apology, however, stressed that earlier accounts were incomplete and made no reference to the valvular folds placed near to the orifice of the pouch.

The ventricle and saccule

Victor Negus when discussing the role of the larynx in preventing air entering the lungs during specific periods of physical activity, described the presence of a secondary valve formed by the inferior thyroarytenoid folds. This was present in those species that used their forelimbs for grasping, hugging, clinging or climbing and this division of the thyroarytenoid muscle resulted in the formation of a lateral recess or ventricle (Negus, 1949). Examination of a large number of species by serial section or macrodissection confirms that although a slit-like depression can be found in the camel (*Camelus bactrianus* and *C. dromedarius*), llama (*Lama glama*), kudu (*Boselaphus tragocamelus*), warthog (*Phacochoerus aethiopicus*), muntjac (*Muntiacus muntjac* and *M. inermis*) and sitatunga (*Tragelaphus spekii*), it is only membranous and covers an undivided thyroarytenoid muscle. Similar findings are found in the jaguar (*Panthera onca*), tiger (*Panthera tigris*) and giant panda (*Ailuropoda elanoleuca*).

The true laryngeal ventricle is lined with pseudostratified ciliated columnar epithelium with a lamina propria consisting of a layer of loose elastic and collagenous fibres with variable numbers of seromucinous glands. Although the size varies considerably with individual species, it is always proportionally larger in the young, as is the saccule (Figure 4.19).

The musculature surrounding the ventricle and saccule varies according to the level of the section although the thyroarytenoid always forms the lateral and inferolateral boundaries throughout the entire length in humans and other primates. The main muscular element of the para-ventricular area is formed by this muscle although variable contributions are seen from thyroepiglottis, superior thyroarytenoid and ventricularis. These muscles, however, are small and variable in presence, size and orientation, probably playing little part in laryngeal function. (Kotby *et al.*, 1991). The difficulty of identifying the superior thyroarytenoid muscle on gross dissection led Santorini (1724) to deny its existence and Zemlin *et al.* (1984) to find it in only 40% of their human specimens. Serial sectioning, however, allows more comprehensive identification and the muscle was found in 80% of primate larynges and to a lesser extent in other species (Figure 4.20). Although an inlet-valve type of larynx with divided thyroarytenoideus muscle and ventricle is present in all mammals with independent action of the forelimbs, the same features are also found in other species that do not normally climb but can dig. Bandicoot (*Metachirus nudicaudatus* and *Didelphis virginiana*), coypu (*Myocastor coypus*), agouti (*Dasyprocta leoporina*), marmot (*Marmota marmota*) coati (*Nasua nasua*), fennec fox (*Vulpes zerda*), wolf (*Canis lupus*), hunting dog (*Lycaon pictus*), badger (*Meles*

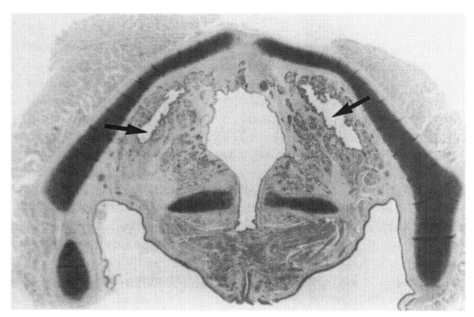

Figure 4.19. Transverse section of the larynx of a six-month-old human showing considerable amount of glandular tissue in the supraglottis and around the large saccules (arrowed).

meles), mongoose (*Herpestes fuscus* and *Khneumia albicauda*) and armadillo (*Dasypus novemcinctus*) all have well formed ventricles. The value of an inlet-valve type of larynx is less obvious in species such as the elephant, or in the Perissodactyla (Figure 4.21).

Embryologically the ventricle and saccule develop as secondary outpouching from the laryngeal lumen around the second month (crown–rump 22 to 24 mm). Neither are vestiges of the visceral pharyngeal pouches. Before the endoderm and mesoderm of the 'chordal nodule' differentiate into the true and false cords, a ventriculo-saccular outpouching progresses laterally from the laryngeal lumen. This outpouching then splits the mesoderm, eventually differentiating into the thyroarytenoid muscles (Hast, 1970) (Figure 4.22).

The saccule arises vertically from the anterior end of the roof of the ventricle and in humans extends superiorly, curving backwards laterally to the false cord and aryepiglottic fold. The relation of the saccular opening to the ventricle has been controversial for although the opening is said to be only a few millimetres in diameter there has been doubt as to the existence of ventriculo-saccular folds. Hilton (1837) described the orifice as being 'guarded' by two crescentic folds of mucous membrane, whilst Delahunty & Cherry (1969) and Kotby *et al.* (1991)

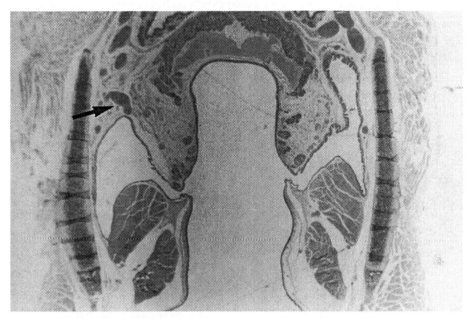

Figure 4.20. Coronal section of the larynx of an agouti (*Dasyprocta leporina*) showing the superior thyroarytenoid muscle (arrowed).

found only a single anterior crescentic mucosal fold passing from the false cord to the lateral wall of the saccule (Figure 4.23). All thought that the purpose of a ventriculo-saccular fold was to help in the storage of mucous and its direction onto the vocal cords.

The histological appearance of the human saccule shows an almost villiform mucosal pattern particularly in the young, with submucosal areas of lymphoid tissue and large numbers of associated seromucinous glands. As age increases the lymphoid tissue decreases but the submucosal lymphatics and glands provide a potential pathway for the spread of glottic cancer to the pre-epiglottic space (Delahunty & Cherry, 1969).

Morphological studies have been carried out with the purpose of determining the dimensions of the laryngeal saccule relative to the height of the larynx in male and female humans. These provide data for diagnosing 'enlarged' saccules and possibly the frequency of undiagnosed laryngocoeles. Scott (1976) carried out a range of measurements in 111 normal larynges finding absence of the saccule in five specimens. Of the male saccules, 70% measured between 0.5 and 0.9 cm in height whilst 78% of female saccules were within this range. The greatest heights obtained were 3.7 cm in a male and 2.15 cm in a female. These

were the only instances where the saccule extended between the upper border of
the thyroid cartilage and the hyoid bone (8%). There was no evidence that
the height of the saccule relative to the height of the larynx differed in the two
sexes.

Birt (1987) carried out similar measurements on 353 normal and neoplastic

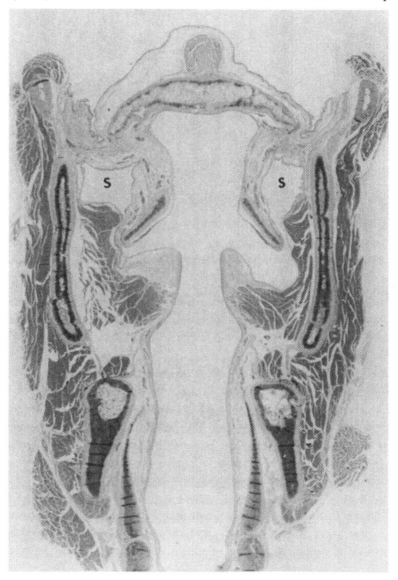

Figure 4.21. Coronal section of the larynx of a wolf (*Canis lupus*) showing bilateral
ventricles with openings into dilated saccules (S) ?air sacs.

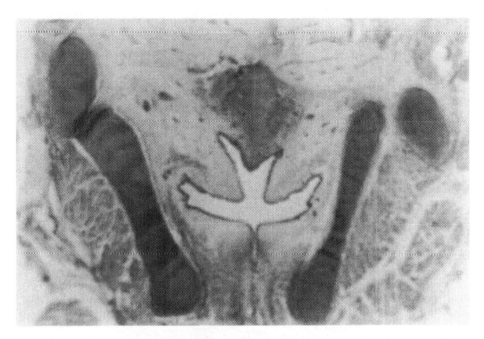

Figure 4.22. Foetal human larynx (aged 22 weeks) cut transversely in the region of the anterior commissure showing embryonic epiglottis ventricles and vocal cords.

human larynges sectioned coronally or horizontally. While he accepted that fixation might have influenced the accuracy of measurements, he carried out multivariate discriminant analysis to determine the percentage accuracy of his findings. In 19 larynges no saccule was identified and in 28 others only one saccule was found. The mean height of the saccules was 8 mm with a range of 3 and 16.5 mm with no sex differentiation; 10% exceeded 15 mm in height. He considered a saccule to be abnormally large if its vertical height exceeded the upper border of the thyroid cartilage or if it extended more posteriorly than the thyroid alar when viewed in horizontal sections. Using this definition Birt (1987) found enlarged saccules, which he called 'laryngocoeles', in 17.8% of this sample. Broyles (1959) examining larynges removed at autopsy, found that 7% of saccules in normal larynges were 1.5 cm or greater in height, with a male preponderance of 3:2. He considered these to be enlarged.

Studies (Harrison, 1983) on 140 serially sectioned larynges removed for carcinoma gave measurements for saccular height similar to those reported by Birt (1987) but also suffered from the possible effects of fixation distortion. However, by using the upper border of the thyroid cartilage as the focal point for establishing an 'enlarged' saccule, variations in measurement techniques can be minimized. A figure of between 7% and 10% would appear to be realistic for the human larynx. There is great variation in the size of the human saccule and the 'guarding' of the opening of the saccule into the ventricle was not seen in every specimen.

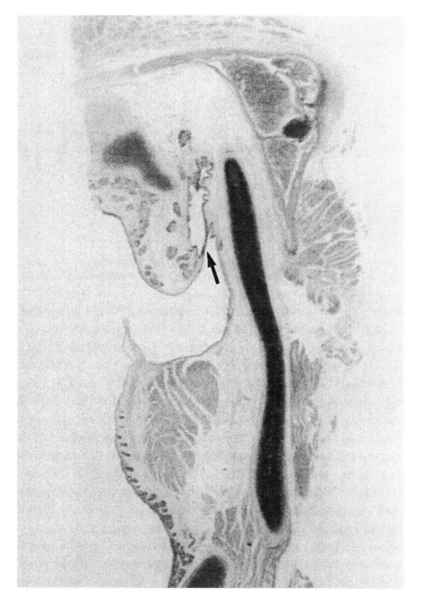

Figure 4.23. Coronal section of half the larynx of a capuchin monkey (*Cebus apella*) showing the mucosal fold (arrowed) 'guarding' the entrance to the saccule.

Comparative studies of the saccule

Ventricles were examined by serial section or macrodissection to determine the presence, size and configuration of the saccule. No foetal specimens were available but serially sectioned immature larynges were studied for 12 species of primate. Despite the presence of a ventricle no evidence of a saccule could be found in the sloth (*Bradypus variegatus* and *Choloepus didactylus*), armadillo (*Chaetophractus villosus* and *Dasypus novemcinctus*), badger, coypu, coati or any of the Perissodactyla, except the horse. The saccule was difficult to identify in some primates such as the lemurs and bush baby (*Otelemur crassicaudatus* and *Galago senegalensis*) and others with ventricular air sacs. This is discussed in detail below. Even when the saccule was clearly identified the presence of associated seromucinous glands was rare, as was lymphoid tissue or a ventriculo-saccular valve although in small larynges the latter could easily be missed (Figure 4.24).

Since many mammals possess large numbers of seromucinous glands within the epiglottis or aryepiglottic folds, it appears that a peri-saccular contribution is relatively unimportant in providing 'lubrication' to the vocal cords. Even when a saccule appeared to be present, the ventricular opening may be large and the sac distended, making differentiation from an air sac difficult (Figure 4.25). All primates have ventricles, and particularly in young animals the saccule may shows features similar to that found in humans. However, the saccule was absent in all the lemurs studied, and in many other adult species it was difficult to differentiate from a ventricular air sac.

Air sacs

Although air sacs in mammals are commonly found in relation to the larynx, they are common in a wide variety of animals ranging from the buccal air sac of the frog to the tracheal diverticulae in the snake (Young, 1992). They are particularly prominent in the siamang (*Hylobates syndactycus*), chimpanzee (*Pan troglodytes*), orang-utan (*Pongo pygmaeus*) and gorilla (*Gorilla gorilla*) (Stark & Schneider, 1960). In the latter two species, museum specimens are said to show that the extralaryngeal sacs meet in the midline beneath the pectoral muscles and may even extend into the axillae. When collapsed, the true capacity of these sacs is difficult to quantify but the largest in relation to body size appear to be found in the apes. The exception is the plains viscacha (*Lagostmus maximus*), a robust rodent with a body length of 60 cm living underground in a complex of burrows in South America. It is of interest that this rodent closes its nostrils to avoid ingestion of soil and may utilize these large subhyoid sacs for respiration

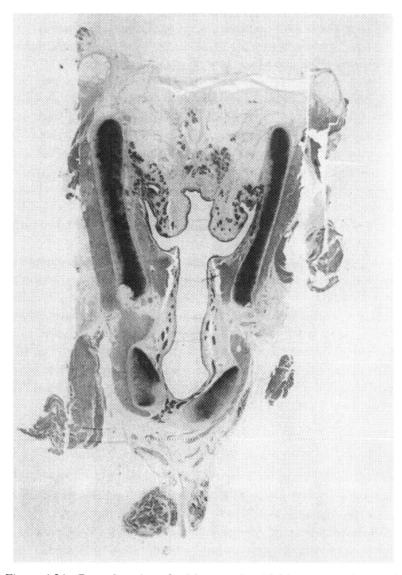

Figure 4.24. Coronal section of colobus monkey (*Colobus satanus*) showing the saccule but no obvious ventriculo-saccular fold.

(Figure 4.26). Although large ventricular air sacs extending between the upper border of the thyroid cartilage and hyoid bone can be clearly identified in adult apes, smaller sacs extending from the ventricle in other species cannot always be differentiated from expanded saccules. This may be due to intrinsic variations

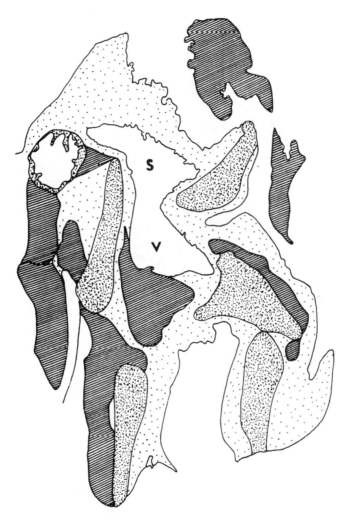

Figure 4.25. Drawing of a sagittal section of the hog badger larynx (*Arctonyx collaris*) showing a large ventriculo-saccular complex (VS). Should this be called an air sac?

within species (as is found in man), errors in determining the age of the animal or inadequate sectioning. Large saccules were found in many primates including bush baby, marmosets, tamarins, macaques, saki, vervet and aotus and capuchin monkeys, mandrill and gelade baboon (Figure 4.27) The relatively close ancestral relationship of the primates may offer some explanation for this finding but a similar situation is also found in the wolf, badger, agouti and bandicoot (Figure 4.28).

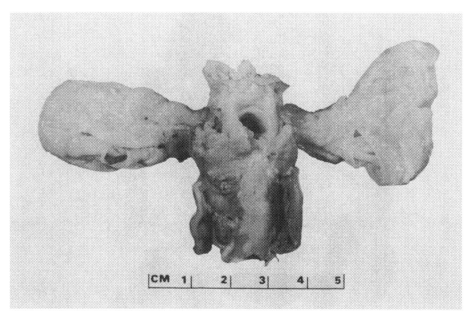

Figure 4.26. Possibly the largest of all laryngeal air sacs per body weight, found in the plains viscacha (*Lagostomus maximus*).

Although the largest of the laryngeal air sacs arise from the ventricle, probably occupying the site of the original saccule, small lateral recesses lying below the aryepiglottic fold, medial to the upper part of the thyroid cartilage are present in the otter (*Lutra canadensis*), collared peccary (*Tayassu tajacu*) and margay (*Felis wiedii*). Other laryngeal sacs have their entrance at the inferior margin of the epiglottis extending into the subhyoid area. In the reindeer (*Rangifer tarandus*) this is usually quite small, serving no apparent purpose, whereas in the plains viscachas (*Lagostomus maximus*) these sacs may assist in respiration.

Air sacs in the chiroptera

Denny (1976) in her research into the anatomy of the bat larynx based on transverse sections of 15 species of microchiroptera and three species of megachiroptera, found upper tracheal air sacs in many specimens. In the Emballonuroidea, the most primitive superfamily, all species have well defined ventricles without a saccule. Glands, however, are plentiful within the epiglottis.

In the Rhinolophidae, considered as being somewhere betweeen the more primitive Emballonuroidae and more advanced Vespertilioninae, the larynx is

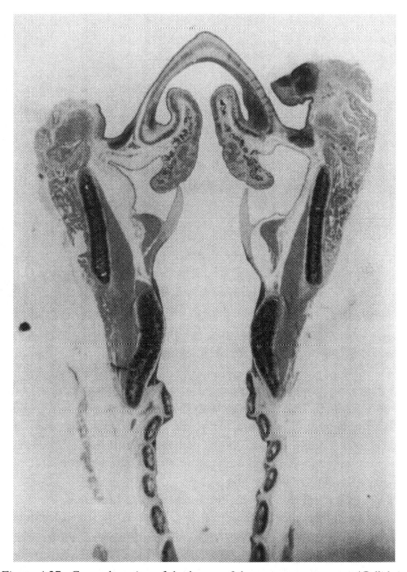

Figure 4.27. Coronal section of the larynx of the common marmoset (*Callithrix jacchus*) showing an expanded saccule (?air sac) which extends above the superior border of the thyroid cartilages.

far more unusual. Apart from its intranarial position and well developed musculature, considered in detail elsewhere, the most important feature are the tracheal air sacs or pouches. Laryngeal air sacs are not found in the chiroptera, but in *Rhinolophus* the first and second tracheal rings are expanded to form a pair

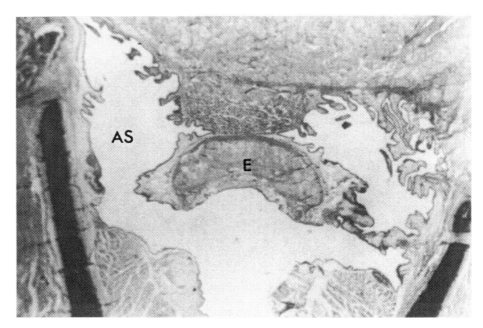

Figure 4.28. Transverse section of the badger (*Meles meles*) larynx showing the
epiglottis (E) and air sacs (AS) extending extralaryngeally.

of lateral pouches, whilst the third and fourth rings are fused dorsally to form a
broad dorsal pouch. The rings fuse with the body of the pouches laterally and
dorsally. Similarly in *Hipposideros caffer*, the cartilaginous rings have also expanded to form a dorsal and two lateral pouches (Figure 4.29). Differences are
seen in *Nycteria* in that the rings are incomplete dorsally with two large pouches
just posterior to the cricoid cartilage.

The Megadermatidae possess no tracheal sacs but small ventricles are
present in *Lavia frona and Cardioderma cor* (Figure 4.30). A similar situation is
found in Phyllostomus, Glossophaga and Desmodus, whilst *Pteronotus
rubiginosa* has a single sac below the cricoid cartilage.

The Vespertilionoidea, considered to be the most advanced superfamily of
the bats and palaeontologically the most recently evolved, shows similarities with
the larynges of the more primitive Emballonuroidea. No tracheal air sacs or
pouches have been found similar to those of the Rhinolophidae, Hipposideridae
or Nycteridae. All have laryngeal ventricles but in *Epomops franqueta strepitens*
there is an anterior extension of both ventricles to produce a large saccule.
Denny (1976) has considered the significance of the tracheal air sacs found
in the Rhinolophidae, Hipposideridae, Nycteridae and *Pteronotus* sp. of the

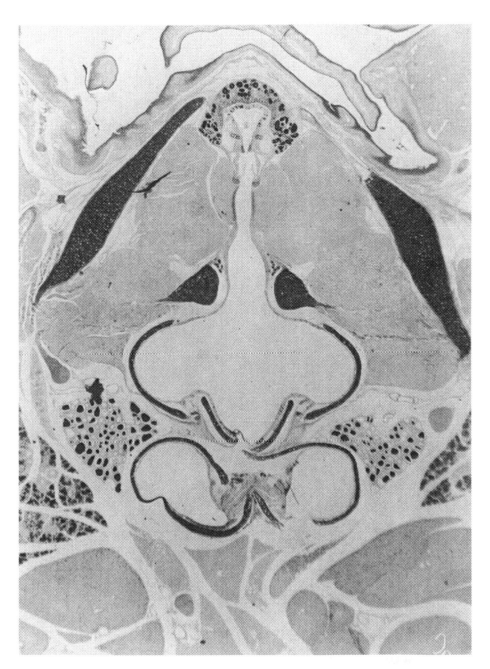

Figure 4.29. Transverse section of the larynx of a leaf-nosed bat (*Hipposideros caffer*) showing posterior expansion of upper tracheal rings. Epiglottis shows many glands.

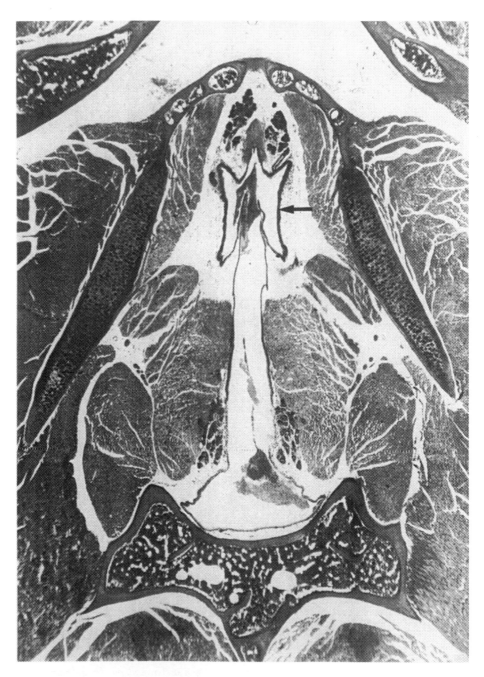

Figure 4.30. Transverse section of the larynx of heart-nosed bat (*Cardioderma cor*) showing the paired laryngeal ventricles (arrowed).

Phyllostomatoidea. These bats are all constant frequency pulse-emitters and with the sole exception of *Pteronotus*, nasal emitters with an intranarial larynx. The latter is an enigma since it is the only member of the superfamily to possess tracheal air sacs, produce constant-frequency pulses and yet is a mouth emitter. It is possible that the tracheal air sacs of some species of bats serve as resonators but certainly they they play no part in aerial respiration.

Function of the primate air sac

There is no evidence to suggest that air sacs are related to vocal intensity, for in many species, such as the lemurs and gibbons, the noisiest animals have no air sacs. Humans are no exception to this and some of the noisiest mammals, such as lion and hyaena possess neither ventricle nor air sac. Negus (1949) offered an explanation for the development of a divided thyroarytenoideus muscle and consequently, a ventricle and double inlet-valve within the larynx, as necessary for fixation of the forelimbs. He felt that this was of particular value in the more active arboreal species of primates and extrapolated his reasoning to the presence of large extra-laryngeal air sacs particularly in the great apes and the siamang. These are classified as modified brachiators although occasionally showing the pure form of arm-swinging locomotion. Anatomically, great apes have the adaptations of brachiators (long arms, hook-like hands, wide chests, etc.) but not the behaviour. The chimpanzee and gorilla are only partially arboreal, whilst the orang-utan spends only 15% of its time on the ground. However, the latter is quadrumanal, using both fore- and hindlimbs to suspend its heavy body (Napier & Napier, 1985).

The adult chimpanzee weighs about 40 kg with a lung volume of one litre, whilst an adult gorilla has a body weight of over 140 kg and an estimated lung volume of at least 10 litres (Schmidt-Nielsen, 1984). If the air sacs, lying beneath the large pectoral muscles of these mammals were to play a significant role in respiration, their capacity would have to be considerable. There is no evidence to substantiate this and moreover, the most active of arboreal primates have at best only small insignificant air sacs, or even none at all. Personal inquiries amongst veterinary pathologists and experienced keepers of great apes reveal little confirmation of the existence of subpectoral air sacs, and these may only be a rare occurrence (Figure 4.31).

The traditional classification of orang-utan, gorillas and chimpanzees in the family Pongidae, and of humans and their fossil relatives in the family Hominidae is now generally accepted (Martin, 1990). Whether or not the completely terrestrial existence of humans is enough to explain the absence of an air sac remains problematic, the protagonists of the 'aquatic' theory of human evolution

Figure 4.31. Adult gorilla photographed at the Jersey Wildlife Park illustrating the difficulty of identifying cervical or subpectoral air sacs in the living animal.

pay no attention to the possible value of an air sac as an accessory lung (Morgan, 1990). In humans, the saccule is probably the vestigial remains of the air sacs which in our closest relatives, the great apes, appear to be relatively functionless.

Laryngocoele

It is generally accepted that the first description of a laryngocoele was by Larrey (1829), a surgeon in Napoleon's army in Egypt. 'It was in Egypt that we first observed examples of this kind of goitre. It showed itself only in one class of individuals, the blind, who are so are so numerous that they are employed to chant verses from the Koran from the tops of the minarets every hour of the day and night. These air tumours chiefly develop in those who have "called the hours" for many years, and produce pockets under the jaws. In order to use their voices, they are obliged to bandage their necks. When the swellings become as large as fists and their voice hoarse, they are retired and are employed minding the temple pools'.

He called these swellings 'goitres aeriens', but in 1867 Virchow introduced the term 'laryngocoele ventriculatis' considering them to be dilatation of the ventricle of Morgagni. They are now considered to be dilatations of the saccule with an incidence according to Stell & Maran (1975) of 1 per 2,500,000 per annum in Great Britain. Laryngocoeles are lined by columnar ciliated epithelium with occasional patches of squamous metaplasia being identical to the ventricular air sacs found in many primates. Unfortunately no entirely satisfactory definition of the human laryngocoele has been reached, with some clinicians describing any saccule which radiologically reaches the upper border of the thyroid cartilage as a laryngocoele. Since this may occur naturally and all saccules contain air, only those which are extra-laryngeal or produce symptoms

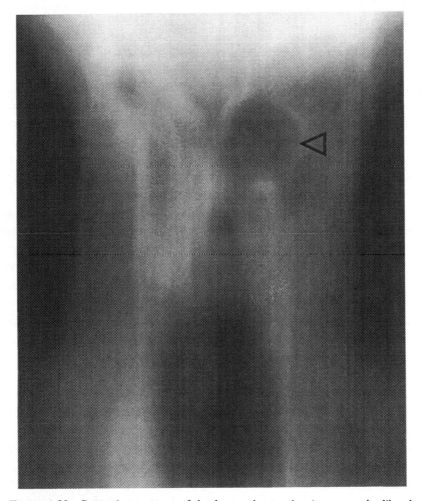

Figure 4.32. Coronal tomogram of the human larynx showing a grossly dilated saccule–laryngocoele (arrowed).

of hoarseness, aphonia or dyspnoea, warrant the diagnosis of laryngocoele (De Santo, 1974) (Figure 4.32).

Despite the experiences of Larrey (1829), the majority of clinically diagnosed laryngocoeles are not associated with excessive blowing, such as in trumpeters or glass blowers. However, raised intraglottic pressure in the presence of an unusually large or long saccule may be still be a factor since the ventriculo-saccular valve cannot be considered as resistant to prolonged raised ventricular pressure. Although morphologically identical to the air sacs found in the great apes, the human laryngocoele appears to be acquired but functionless, telling us little about our inheritance.

4.4 Laryngeal mucous glands

That part of the laryngeal cavity lined by ciliated, columnar respiratory-type epithelium has varying amounts of submucosal mucous glands. These are of the exocrine compound tubulo-alveolar type, each alveolus being a cluster of cells making up a single secretory unit. Characteristically, large glands have extensive duct branching systems. The main duct branches into ever smaller and more numerous branches that drain the secretory units of a compound gland. These seromucinous glands may be embedded in a fibroareolar stroma within the false cords admixed with muscle fibres. All laryngeal glands are of the mixed seromucinous type and being exocrine in function share the general histological features of such glands. The acini and ducts have a dual cell population – an outer myoepithelial layer of cells and an inner luminal epithelial secretory cell (Nassar & Bridger, 1971). Many anatomy textbooks quote a total of around 70 glands within the larynx, although it is not clear whether this represents an average of many observations. Counting of individual glands is technically difficult; Harrison (1983) in his observations of 141 serially sectioned human larynges showed great variation in the number and siting of all the laryngeal glands.

Although they did not attempt to count the number of glands, Nassar & Bridger (1971) examined a small number of normal human larynges sectioned in the horizontal or coronal plane, with a view to determining the normal distribution of laryngeal glandular tissue. They divided the larynx into three topographical regions – glottic, supraglottic and subglottic. Within the supraglottis, the epiglottis in humans contains many glands in the inferior two-thirds of the laryngeal surface. They usually lie within the many pits that pass through the epiglottic cartilage and are also within the aerolar tissue of the pre-epiglottic space. At the lowest portion of the epiglottis the glands extend to the anterior commissure and adjacent subglottis. This group Nasser & Bridger term the medial group. Laterally, in proximity to the thyroid alar many glands are associated with the laryngeal saccule and these can extend into the pre-epiglottic space. They are separated by fibres of the thyroepiglottic and thyroarytenoid muscles from the central group, and these glands are called the 'lateral' group (Figure 4.33).

The aryepiglottic folds are composed of loose aerolar tissue and usually contain few glands, which is not surprising in view of the covering of squamous epithelium. The epithelium of the false cords is of the respiratory type although varying amounts of squamous metaplasia may occur in males (Stell *et al.*, 1981). Many glands are usually present in the submucosa and within the quadrangular membrane. They are separated from the more laterally placed 'lateral saccular' glands by aerolar tissue and fibres from the vocalis muscle. More anteriorly they join with the medial group of epiglottic glands (Figure 4.34). As has been shown in section 4.3 devoted to the ventricle and saccule, the density of glandular tissue in this area depends upon the size of the saccule. Glands may also be found in the paraglottic region. Posteriorly, glands are invariably aggregated around the cuneiform and corniculate cartilages extending to the ventrolateral surface of the arytenoid. At the posterior commissure the epithelium is also respiratory in type. Abundant seromucinous glands are present and may extend deeply into the interarytenoid muscle.

The glottis and subglottis

Squamous epithelium is usually confined to the free edge of the vocal cord although some extension to the floor of the ventricle or into the subglottis was found by Scott (1976) to occur in 11% of 111 normal larynges. The lamina propia of the vocal cord contains seromucinous glands but these are found in the submucosa of the true cord only rarely. Subglottically, most glands are found within the subepithelial elastic lamina and the ducts may extend submucosally down to the conus elasticus.

It is generally accepted that the prime function of the laryngeal mucous glands is lubrication of the respiratory-type epithelium. In addition, the stratified squamous epithelium covering the edges of the vocal cords is vulnerable to drying and saccular secretion has been observed to flow over the cords during phonation.

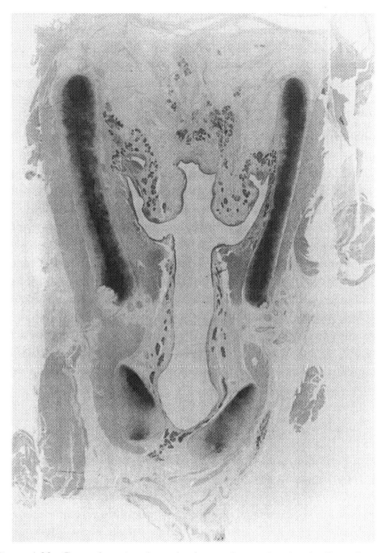

Figure 4.33. Coronal section through a human larynx showing the 'lateral' group of glands in relation to the saccule.

Possible role of laryngeal glands in the spread of cancer

When discussing carcinoma *in situ*, Altmann *et al.* (1952) observed that carcinoma of the larynx involving areas into which excretory glands opened, tended to grow into these ducts. Involvement of duct epithelium would allow

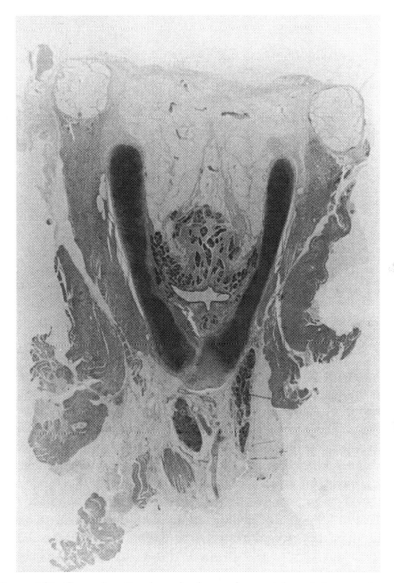

Figure 4.34. Coronal section through a human larynx at the anterior commissure showing both lateral and medial group of glands forming a single mass.

non-invasive intra-epithelial cancer to spread beneath the surface epithelium. Bridger & Nassar (1971) examined two serially sectioned larynges where there had been extensive glandular involvement by *in situ* carcinoma. They found carcinoma within glands some distance from the mucosal opening of the ducts,

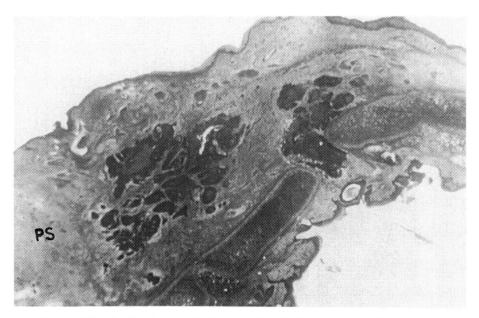

Figure 4.35. Section through the human epiglottis showing cancer within one pit and extending anteriorly to the pre-epiglottic space (PS).

concluding that connective tissue barriers within the larynx offer little resistance to the spread of ductal cancer.

Squamous metaplasia of the ciliated columnar epithelium in the larynx and encroachment of this metaplastic epithelium into the glandular ducts is relatively common. Glandular involvement by *in situ* carcinoma probably arises in a similar manner, and within the supraglottis may be responsible for unexpected, and undiagnosed, involvement of the pre-epiglottic space.

In 1983 Harrison reported his findings on 141 serially sectioned larynges removed for carcinoma. He emphasized that the sero-mucinous gland, which occupied pits on the laryngeal surface of the epiglottis, offered easy access to the pre-epiglottic space for even small cancers (Figure 4.35). Many tumours that on clinical and radiological grounds would be designated T1 or T2, were really T4 and carried a poor prognosis (Harrison, 1983).

The immature larynx

In his studies on the immature mammalian larynx, Pracy (1984) found that significant morphological differences occurred between the immature and adult larynx in many mammals. These he felt were primarily related to the suckling

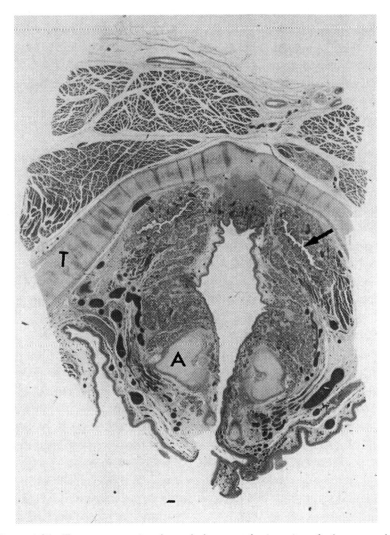

Figure 4.36. Transverse section through the supraglottic region of a three-month-old girl to show dense accumulation of glandular tissue. Arytenoid (A), thyroid cartilage (T), saccule arrowed.

phase of life for all mammals, irrespective of their later behaviour pattern, drink milk when young. These structural differences are discussed elsewhere but in many immature mammals the mucous secreting glands appear to be more active than in the adult and widely distributed (Figure 4.36). Why should the baby produce more mucus when extra lubrication of the larynx and upper alimentary tract appears unnecessary? Most of these glands are found within the supraglottis and Pracy (1984) has suggested that since both human and some other mammalian neonates have airway difficulty at birth, this mucus may form a temporary 'seal' to prevent inhalation of amniotic fluid. The bulky distension of these glands forms an actual increase in the soft tissue volume of the supraglottis and whilst supporting the aryepiglottic folds may also produce reduction in the laryngeal lumen.

Sudden infant death syndrome (SIDS)

The high posterior laryngeal wall, bulky aryepiglottic folds and large cuneiform cartilages common to the human infant larynx protect the respiratory tract from ingestion of milk. Although most of the sero-mucinous glands are in the supraglottis, glands are also present within the subglottis, glottis and extending to the lower border of the cricoid cartilage (Figure 4.37). The possibility of this being a factor in the aetiology of SIDS has been investigated by Harrison in 1991.

Before SIDS was classified as a specific entity in the eighth revision of the *International Classification of Diseases* in 1970, these deaths were recorded under the heading of respiratory diseases. Much of the fall in infant mortality since this date has been due, in the United Kingdom, to a gradual reduction in neonatal rather than post-natal mortality. Since the definition of SIDS – 'sudden unexpected death where after a thorough post-mortem examination no cause for death can be found' – is open to misdiagnosis, the true incidence is uncertain. However, SIDS occurs particularly in the Western world where it is common for young babies to sleep alone in a cot at night. This is a practice which was uncommon before the seventeenth century, and appears to have been promoted in England by the Victorians for convenience (Gregson, 1989). Although the sleeping position appears to be a factor in these tragedies, SIDS appears to be most common in the winter, two-thirds are in babies between the age of two and six months and signs of minor upper respiratory infection are a feature of many reports. In 1980 Fink & Beckwith examined the larynges of 13 SIDS babies together with 10 infants dying from known causes. Following transverse sectioning, measurements were made of the areas occupied by mucous glands and a 'mucous gland index' calculated for the supraglottis. The mean index for the 13 SIDS larynges was 19.52 ± 8.56, and for the controls, 12.56 ± 6.25, the difference being said to be statistically significance. They admitted that obtaining matched controls that had not been intubated or had had other laryngeal trauma was difficult and concluded that the proportion of mucous glandular tissue in the vestibular folds of SIDS victims was twice as great as in the control group. They suggested too, that it might be the amount of secretion rather than number of glands that was the important feature in the

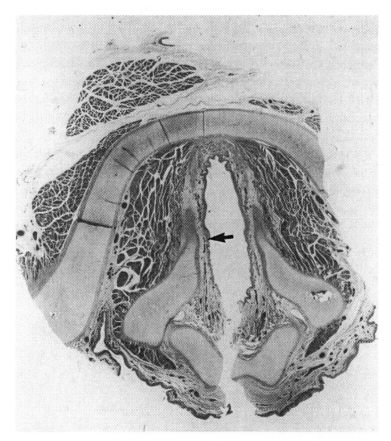

Figure 4.37. Transverse section of normal larynx at the level of the vocal cords from a four-month-old human showing minimal amount of subepithelial glandular tissue (arrowed).

aetiology of SIDS. Increase in quantity and perhaps viscosity in small larynges could produce undetected respiratory embarrassment. Harrison (1991) studied 104 larynges registered as dying from SIDS. He felt that serious reduction in laryngeal airway was more likely to occur in the subglottis; the rigid confines of the cricoid cartilage would maximize the effect of glandular hypertrophy and the consequent reduction in subglottic lumen. After embedding the larynges in paraffin wax, serial sections were cut at 6 μm in the transverse plane. Each section was then measured using a Measuremouse Imaging System (Analytical Measuring System, Cambridge, England) which allows accurate reproducible measurements of a wide variety of parameters, including area. Images projected under magnification onto a visual display unit can then be measured using a tracking device to outline desired features. The glandular tissue within the supraglottis was measured in 10 SIDS and an equal number of control larynges without confirming the variations found by Fink & Beckwith (1980). The gross amount of glandular tissue in this region in normal infants and the technical difficulties in accurate measurement made this an unsatisfactory investigation. Even if there were significant differences it is doubtful if the airway would be seriously compromised. Measurements of the cricoid lumen (the theoretically available area for respiration) and the area occupied by glandular tissue, was made in all cases (Figure 4.38). Some reduction in available area was present in every larynx, since some submucosal glandular tissue was always present. No data is available relating to the minimal subglottic area necessary for normal

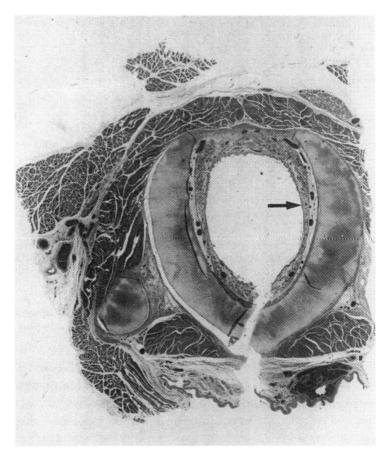

Figure 4.38. Transverse section at subglottic level from a three-month-old boy dying from congenital heart disease, showing moderate amount of submucosal glands (arrowed).

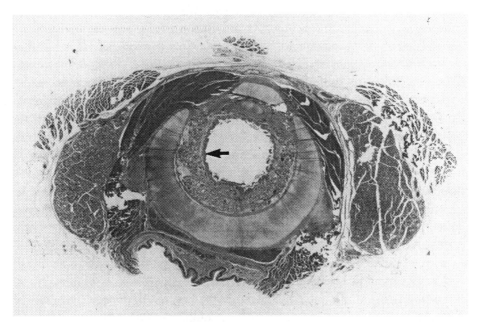

Figure 4.39. Gross reduction in subglottic lumen in a three-month-old boy dying from SIDS. Note the dramatic increase in subepithelial glandular tissue (arrowed).

respiration in this age group, and the measurements from the 10 control larynges were taken as the normal baseline. A mean figure of 38.5% of available area was found for the two- to five-month-old babies. However, in view of the difficulty in obtaining control larynges a figure of 50% reduction in available airway was taken as the 'normal' baseline when analysing the SIDS larynges.

No infant older than four months showed more than a 45% reduction in subglottic airway but 35% of the larynges aged two to four months had a reduction of more than 50% of their airway; 30% had a reduction greater than 60% (Figure 4.39). In one three-month-old boy weighing 3300 g at birth and with a death weight of 3430 g, the subglottic area was reduced from a possible 26.62 mm^2 to only 9.12 mm^2. There was no statistical evidence that maximum subglottic area was directly related to age or body weight at birth or death.

These findings suggest that in 30% of babies within the ages of two and four months registered as dying suddenly without a discernible cause, the subglottic airway was reduced by as much as 50–60%. Further research on the presence of sulphanated and carboxyllated sialomucins within the glandular tissue responsible for this reduction in subglottic airway may confirm a suspicion that there is also an increase in mucin viscosity.

Comparative studies

Histological examination has provided valuable information about the distribution and activity of the submucosal glands which form a prominent feature of the tissue structure of many mammalian larynges. However, as Pracy (1984) found in his morphological studies on immature mammalian larynges, there is considerable variation in disposition between individual species. Quantification of glandular tissue has proved technically difficult but sections suggest an increase in both numbers, and possibly activity, between the immature and adult animal in many species. In almost all immature larynges the mucosa of the arytenoids is prominent and swollen, in some cases this is associated with increased activity of the submucosal mucous glands. Pracy measured the ratio between the height of the arytenoid complex relative to the height of the cricoid lamina in the same plane. In 70 larynges, carnivores had the lowest posterior wall, with a mean ratio in the adult of 0.46 reducing to 0.56 in the immature animal. Similar calculations were carried out on Primates, Perissodactyla and Artiodactyla and the significance of these measurements is discussed in the section on morphology of the supraglottis in Chapter 3. It would be helpful if it were possible to demonstrate that engorgement or increase in numbers of the submucosal glands of the immature larynx was consistent. It would then be possible to argue that the functional reason behind such an increase was to support lax tissues in the supraglottis, thus providing support and higher lateral walls as a safeguard against overspill of milk. Although mucous engorgement within glands does provide an increase in support in some species, it is far from universal. Pracy summarized his conclusions regarding the morphology of the mucous glands of the immature human larynx as:

(A) Exceptional prominence of the 'cuneiform' swelling so that it may be mistaken for the corniculate cartilage, whereas it is in fact simply a sausage shaped swelling of glandular tissue. This is tense due to the fluid contained and acts as a supporting tissue to the aryepiglottic fold in the human.
(B) Generalized swelling of the false cord area narrows the laryngeal introitus producing a relative increase in the depth of the glottis
(C) Increase in the volume of the false cords and the amount of glandular tissue at the base of the epiglottis accentuates the curling of the petiole of the epiglottis, producing narrowing of the inlet.

Other immature primates   Although an inability to study the immature larynges of gorilla and chimpanzee limits our knowledge of our closest relatives, other primate larynges show a number of common features. Apart from the

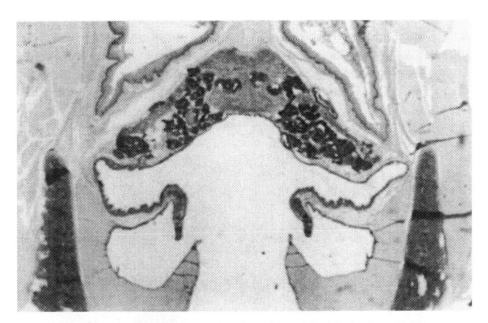

Figure 4.40. Although the laryngeal surface of the epiglottis in the ring-tailed lemur (*Lemur catta*) contains many glands there are only a few actual pits within the epiglottis.

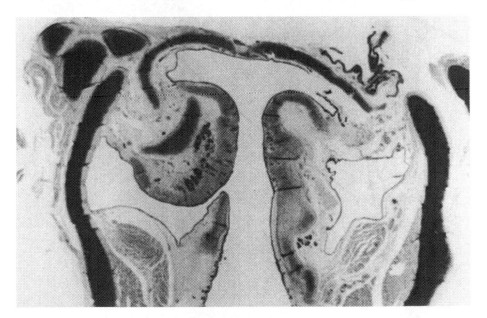

Figure 4.41. In the saki monkey (*Pithecia pithecia*) not only are there no glands associated with the epiglottis, but only a minimal amount are related to the large laryngeal air sacs.

crab-eating macaque (*Macaca irus*) and the sooty mangabey (*Cercocebus atys*) no primate showed more than occasional epiglottic pits (Figures 4.40 and 4.41). All, however, had glands in relation to the inferior aspect of the epiglottis with varying degrees of laterally placed glandular tissue. These were not significant in

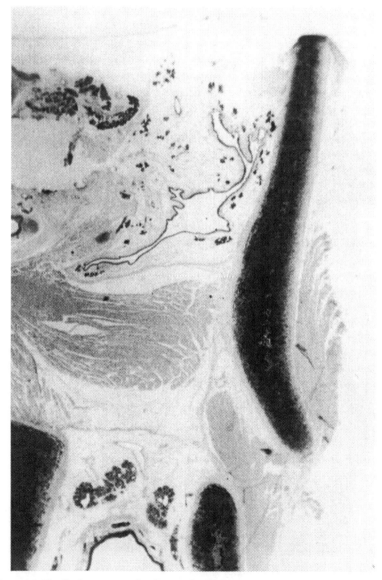

Figure 4.42. Rather more glands are found associated with the saccule in the mandrill (*Mandrillus sphinx*), but these are sparse compared with those of humans.

relation to the air sacs, and only in the lar gibbon (*Hylobates lar*) and mandrill (*Mandrillus sphinx*) were saccules identified and found to be closely associated with mucous glands (Figure 4.42). This suggests that differentiation between an expanded saccule and a functional air sac may be the presence of numerous glands in the former. Most immature primates show glandular tissue in relation to the arytenoid (e.g. orang-utan) although glandular tissue is sparse at all sites in the slow loris (*Nycticebus coucang*) and saki monkey (*Pithecia pithecia*). Presence of glands in relation to the arytenoid is accompanied by varying amounts of glandular tissue in the false cord but considerable variation is seen within individual species and also between individual members of the same species. Apart from the sooty mangabey, few glands were found in the subglottic region and most acini distension occurred within glands related to the arytenoid and false cord.

Immature carnivores   The wide variety of species within this order makes comparative studies difficult but within the 'cats' epiglottic pits are rare and supraglottic glands are uncommon, except in relation to the arytenoid. Most glandular tissue is found in the subglottis. Although even here there is little evidence of glandular engorgement. The coati (*Nasua nasua*) is particularly interesting since it appears to have almost no laryngeal glands (Figure 4.43).

Immature herbivores   Despite the need for high lateral laryngeal walls in both the immature and adult animal, glandular topography shows considerable variation. Whilst the young reindeer (*Rangifer tarandus*) has many epiglottic pits and associated glands in the aryepiglottic folds, the pigmy hippo (*Choeropsis liberiensis*) has no pits and only a few scattered mucous glands. Similar variations can be found amongst other members of the orders Marsupialia, Edentata, Insectivora, Chiroptera, Pinnipedia, Hyracoidea, Lagomorpha, Tubulidentata and Rodentia. However, inability to examine many immature specimens within these orders may have prevented an understanding of the significance of the presence or absence of increased glandular activity in the immature mammal. Even their importance in the human remains questionable.

The adult mammal

Examination of serially sectioned specimens confirms that there is no clear pattern of glandular distribution in adult mammals whether they are classified on zoological or behavioural systems. No marsupial was found to have pits within the epiglottis yet glands were abundant on the laryngeal surface and within the aryepiglottic folds on all species examined, except the kangaroo. These animals make little sound and the purpose of this glandular profusion remains unclear. Amongst the Edentata the epiglottis of the hairy armadillo (*Chaetophractus villosus*) is unusual being almost all pits filled with glands but none are found within the false cord (Figure 4.44), whilst the giant anteater

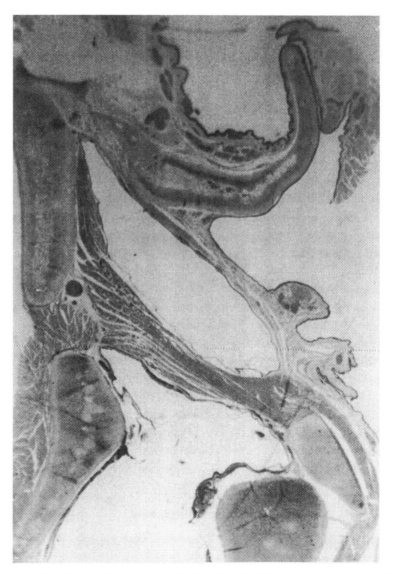

Figure 4.43. No glandular tissue can be identified in this coronal section through the larynx of a baby coati (*Nasua nasua*).

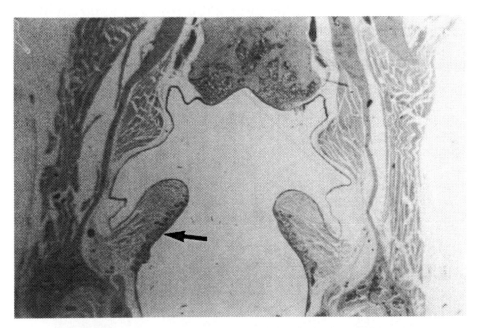

Figure 4.44. The hairy armadillo (*Chaetophractus villosus*) shows another variation in the distribution of glandular tissue. The epiglottis contains many glands but none are found within the false cord or related to the ventricle. Glands are present beneath the true cords (arrowed).

(*Myrmecophaga tridactyla*) has no epiglottic pits and only a few scattered glands within the supraglottis. Similar variations are to be seen within the Primates where epiglottic pits are only seen in the sooty mangabey, macaques, gibbons and chimpanzee. Most primates show glandular tissue at the base of the epiglottis and false cord. In the lemurs this is extensive within the paraglottic region and extends into the pre-epiglottic space, despite the absence of saccule or air sac (Figure 4.45). Glands are not prominent in relation to air sacs although in the Aotus monkey (*Aotus trivirgatus*) they are seen below the vocal cords (Figure 4.46).

Most carnivores have no epiglottic pits although the amount and distribution of glandular tissue varies widely. Whereas the fennec fox (*Vulpes zerda*) has many glands at the base of the epiglottis and around a large saccule, the puma (*Felis concolor*) like most of the large 'cats' has only a scattering of glands, many of which are subglottic. Both the hog badger (*Arctonyx collaris*) and otter (*Lutra lutra* and *L. canadensis*) have many glands in the epiglottic pits and aryepiglottic folds. No clear pattern emerges that explains these variations in glandular topography.

The Rodentia shows similar variations, with the coypu (*Myocastor coypus*) and prairie dog (*Cynomys ludovicianus*) having an epiglottis with many glands (Figure 4.47), whilst the agouti (*Dasprocta leporina*) has minimal glands despite the presence of an enlarged saccule/air sac. Similar variations are seen within the

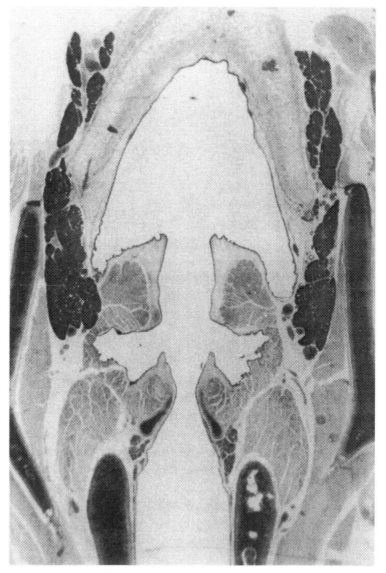

Figure 4.45. Despite an absence of a saccule or air sac the lemurs show extensive glandular tissue in the paraglottic region extending into the pre-epiglottic space. Coronal section is from the ruffed lemur (*Varecia variegata*).

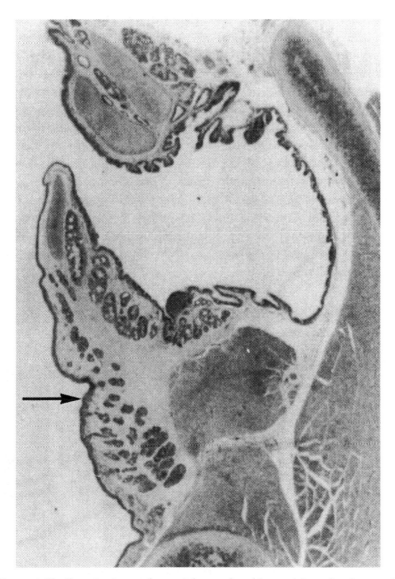

Figure 4.46. Even in the northern night monkey (*Aotus trivirgatus*) only a small amount of glandular tissue is related to the air sac. Glands are, however, found beneath the vocal cords (arrowed).

Perissodactyla, where the epiglottis of the tapir (*Tapirus terrestris*) is almost all glandular tissue extending into the aryepiglottic fold, whilst the onager (*Equus hemionus*) has only minimal glandular tissue within the aryepiglottic fold. In general the herbivores, most being within the order Artiodactyla, have both pits and glands in relation to the epiglottis (Figure 4.48). These usually extend into the aryepiglottic fold and occasionally the subglottis. However, both the amount and distribution is variable between and within individual species. Without studying much larger numbers of specimens it is not possible to achieve more than a general impression of the wide variation which exists within adult mammals in the distribution of laryngeal glandular tissue. The activity of these glands may be determined by factors such as the cause of death, age, sex and other unknown factors. Sufficient to say that this study has emphasized the important role the laryngeal glands may play in the protection of the immature larynx, being possibly a contributory factor in the aetiology of SIDS.

The purpose of laryngeal glands in the mature adult remains obscure and is neither related solely to vocal cord lubrication nor to aiding deglutition. Whether the topography of the glands in the human plays a significant role in the spread of laryngeal cancer remains uncertain but it does present a potential pathway to extralaryngeal tissues.

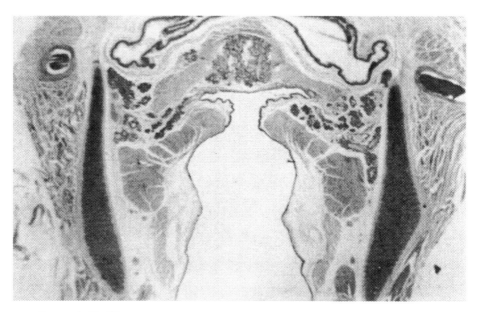

Figure 4.47. The prairie dog (*Cynomys ludovicianus*) has many glands within the epiglottis as well as within the aryepiglottic fold.

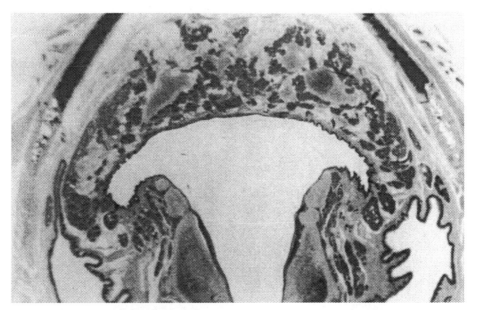

Figure 4.48. In the Chinese water deer (*Hydropotes aquaticus*) the epiglottis is almost completely composed of glandular tissue which extends laterally into the aryepiglottic folds.

4.5 Musculature

The need for a continuous supply of energy is paramount in all mammals. This is provided by oxidation-reduction reactions that themselves depend upon oxygen as the oxidant, and molecules derived from food as the reductant. Fink (1978) emphasized that the rate-limiting factor in the inflow of energy is the flow of oxidant, and in air-breathing mammals the larynx becomes the limiting control point or modulated 'bottle-neck'. The most ancient primordium of a larynx is represented by the sphincter-like structure found at the entrance of the primitive respiratory tract of the lung-fish. The main function of this purely muscular sphincter is to protect the lower respiratory tract. Evolution of this area led to the introduction of cartilages, and the sphincter muscles were divided to subserve many functions (Kotby & Haugen, 1970a,b). The 'bottle-neck' became enlarged by the addition of a thyroarytenoid complex with thyroarytenoid folds, supporting thyroid cartilages and protective epiglottis. The emergence of modulated vocal activity then led to a need for variation in tension of the thyroarytenoid folds, which in turn required synovial cricothyroid joints and cricothyroid muscles (Figure 4.49). The primary and fundamental mechanical control for inflow of oxygen then becomes additionally, the means of controlling emission of information, this is discussed in more detail elsewhere.

The opening and closing of the larynx is determined by extrinsic and intrinsic

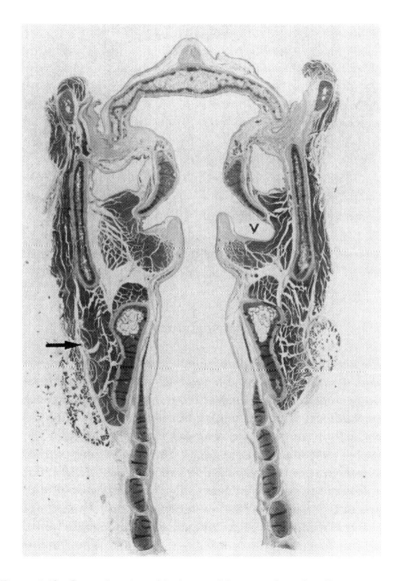

Figure 4.49. Coronal section of the larynx of the squirrel monkey (*Saimiri sciureus*) at the level of the vocal cords to show musculature. Cricothyroid muscle (arrowed), ventricle (V).

muscles acting on the passive mass and elastic forces of the tongue, pharynx, larynx and trachea. There is general agreement regarding the broad lincs of laryngeal function. Opinions differ concerning the precise details by which the laryngeal muscles act on the laryngeal airway, and this is discussed in the sections devoted to laryngeal physiology. The action of the intrinsic muscles depends upon the resultant pull on the cricoarytenoid and cricothyroid joints. This causes tensing of the vocal cords or movement of the arytenoid cartilages. Studies on both muscles and joints have revealed the wide morphological variation to be found within these structures (Konrad *et al.*, 1984). Little research has been carried out on comparative studies of the cricoarytenoid joint and this important subject is considered separately. However, Sellars (1978) and then Kotby *et al.* (1991) have drawn attention to the lack of studies of the precise morphology of the intrinsic laryngeal musculature. By combining data obtained from dissecting fresh and preserved larynges and studying serial sections it has been possible to determine precise origins and attachments and interspecific variations. This work has comprised studies of human and other mammalian larynges, using both macroscopical dissection and examination of more than 250 serially sectioned specimens. Whereas large numbers of human larynges are readily available it is difficult to obtain similar numbers of non-humanoid specimens, except for common laboratory animals, those bred for food or found in herds, such as zebra or deer. Consequently, data relating to the incidence of muscles such as the ceratocricoid or horizontal belly of the cricothyroid are restricted to the study of small numbers of a wide variety of species. Similar restrictions occur with regard to sexual dimorphism, although no evidence is available to suggest that except for possible variations in muscle mass, the anatomy of the laryngeal musculature is influenced by sex.

Posterior cricoarytenoid muscles

These paired triangular shaped muscles are the largest within the larynx originating from the posterior surface of the cricoid lamina either side of a central ridge. The shape of the muscle demonstrates a variation in direction of its fibres, the medial superior of which pass laterally and upwards, the medial inferior laterally and obliquely upwards and the lateral fibres nearly vertically. Sellars (1978) found a lateral origin in 17 of 23 larynges that was attached to the base of the arytenoid and meeting the insertion of the lateral cricoarytenoid muscle. The insertion into the posterior aspect of the muscular process of the arytenoid is well recognized in the human and in most of the other species studied (Figures 4.50 and 4.51). Whether the complex movements of the

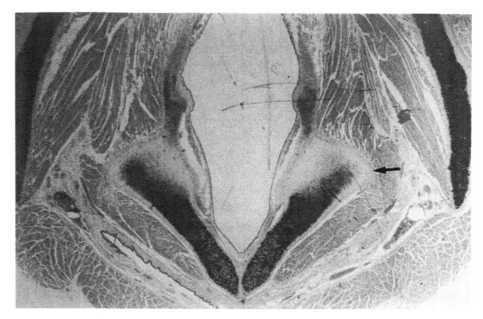

Figure 4.50. Transverse section through the larynx of coati (*Nasua nasua*) showing insertion of the posterior cricoarytenoid muscle into the tip of the muscular process of the arytenoid (arrowed).

cricoarytenoid joints can be explained by the diversity of direction of these muscle fibres and their attachments to the arytenoids is less certain. Variations in the anatomy of the joint within individual species are common, as are marked differences in the shape and size of the muscular process of the arytenoid cartilage (Figure 4.52).

Ceratocricoid muscle

This small and variable muscle called the keratocricoid, ceratocricoid or Merkel's muscle has been known since the time of Vesalius. Although initially described as the second fasciculus of the cricothyroid its distinct nature was described by Merkel in 1857, thirteen years after Tourtual called it the cricocorniculate muscle (Turner, 1860; Sharp, 1990). Many anatomical texts fail to mention this small muscle although Sellars found it to be present in 3.2% of 23 human larynges (Sellars, 1978). Hetherington (1934) found a higher incidence in both the American white (12.9%) and in the African-American (7.6%) populations. A more exhaustive study was carried out by Sharp (1990)

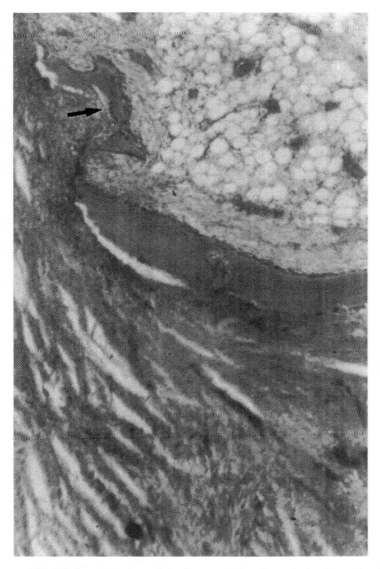

Figure 4.51. High power view of transverse section through the larynx in the
Tasmanian devil (*Sarcophilus harrisii*) showing muscle fibres penetrating the
perichondrium (arrowed).

on 73 adult human larynges. One hundred and thirty-four hemilarynges were
available, the remainder having been damaged prior to the investigation. Nine
ceratocricoid muscles were identified – a prevalence of 6.7%. In two specimens
the muscle was found to be bilateral. In my own investigation of 84 larynges

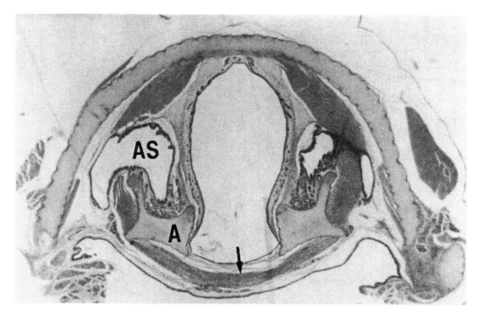

Figure 4.52. Transverse section of the larynx of northern night monkey (*Aotus trivirgatus*) to show air sacs (AS), unusually shaped arytenoids (A) and inter-arytenoideus muscle (arrowed).

removed for neoplasia, six muscles were found giving an incidence of 7.1%, all were bilateral. In each case the ceratocricoid muscle arose inferolaterally or below the posterior cricoarytenoid muscle on the external face of the cricoid cartilage. Passing laterally and slightly upwards it was inserted on the posteromedial side of the inferior horn of the thyroid cartilage. The muscle is always small and partially obscured by the larger posterior cricoarytenoid muscle. Anteriorly lies the superior posterior cricothyroid ligament as it runs superomedially over the cricoid cartilage towards the cricoarytenoid joint.

It has been suggested that this muscle is a weak antagonist of the cricothyroid muscle; this seems unlikely in view of its position and size. Sharp (1990) suggests that it is most probable that it enhances the stability of the cricothyroid joint, and in some animals that it developed as a result of the joint's potential weakness. A search was made for the ceratocricoid in 69 specimens of non-humanoid mammalian larynges dissected to show the morphology of the cricothyroid muscle. (Appendix 2). With cricothyroid muscles less than 4 cm in area, such as in coyote, lynx, coati and rock hyrax, identification was difficult even under magnification. With the sable antelope (17.62 cm) or Cape buffalo (27.0 cm) identification was assisted by the large cricoid plate. However, a ceratocricoid muscle was only identified bilaterally in one specimen of a bison (*Bison bison*). The real incidence of this relatively unimportant muscle awaits more detailed examination of much larger numbers of individual specimens.

Lateral cricoarytenoid

By comparison with the posterior cricoarytenoid, this muscle is small and it appears to be a joint stabilizer rather than directly moving the arytenoid. It arises on the external surface of the side of the cricoid cartilage up to and including its superior border. The fibres lie parallel and passing upwards and backwards are attached at the base of the arytenoid cartilage (Figure 4.53A). As with other muscles attached to the arytenoid cartilage the pattern of attachment at the base of the arytenoid is variable. Mossallam *et al.* (1987) studied this in six young adult larynges using microscopy to define the actual muscle attachments. They measured the extent of each attachment and its relation to the arytenoid surfaces. With this data they constructed a diagram of each surface of the arytenoid with the sites and extent of attached muscle and tendinous fibres. The extension of insertion of the lateral cricoarytenoid above the maximal concavity of the arytenoid facet varied from 200 to 600 μm. The upper limit of insertion extended posteriorly 2200 to 4200 μm beyond the maximal concavity, whilst the lower limit extended 1000 to 1200 μm beyond the maximal concavity in two

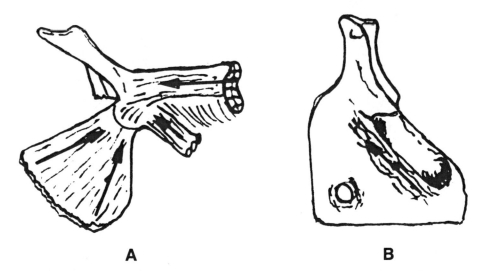

A   B

Figure 4.53. Diagramatic representation of (A) muscular insertions (arrowed) into the arytenoid, (B) the origin of the lateral cricoarytenoid muscle (arrowed) from the external surface of the cricoid cartilage. (After Sellars, 1978.)

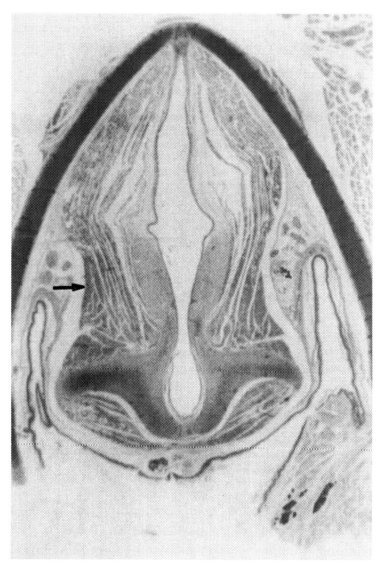

Figure 4.54. Tree hyrax larynx (*Dendrohyrax arboreus*) showing attachments of thyroarytenoideus muscle (arrowed).

specimens but failed to reach the maximal concavity of the arytenoid in others. Similar variations were found for the thyroarytenoid and interarytenoid muscles, confirming the deviations which are a noticeable feature of morphological anatomy.

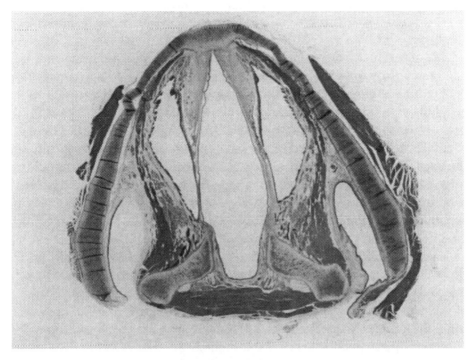

Figure 4.55. Broad anterior attachment of thyroarytenoideus muscle to the thyroid cartilage in a chimpanzee.

Thyroarytenoid and superior thyroarytenoideus muscles

Although often described as a 'broad, thin muscle', the thyroarytenoid is surprisingly large in many species (Figure 4.54). Its origin on the back of the thyroid cartilage is triangular shaped and placed inferiorly on either side of the thyroid angle. Although small in humans, the origin is also large in many other mammals (Figure 4.55). As the muscle passes posteriorly external to the cricovocal membrane, fibres originating from this membrane join it providing additional bulk. Depending on their origin and final insertion on the arytenoid, the direction of the various fibres varies. This is best seen in the bigger non-humanoid larynges with the muscle inserted onto the vocal process of the arytenoid covering its lateral, superior and inferior surfaces. The direction of the different fibres varies – the most superior passing horizontally, the lowest medial fibres obliquely upwards and backwards, and the most lateral fibres join those of the lateral cricothyroid muscle. In those mammals with a laryngeal ventricle the thyroarytenoid muscle forms the inferolateral and lateral boundaries throughout its entire length. The thyroepiglottis on the other hand forms the lateral boundary of the ventricle only in mid-larynx, where the muscle fibres depart from an anteroposterior course to pass vertically joining the lateral margin of the epiglottis.

The superior thyroarytenoideus is another of the muscles occupying the paraventricular area (Kotby *et al.*, 1991). Santorini (1724) and Morgagni (1741) both described this muscle although it is rarely recognized in anatomical texts and difficult to identify in macroscopical dissection. Kotby *et al.* (1991) examining coronal serial sections of 20 normal human larynges found the muscle bilaterally in 80%. It appeared as a round or oval bundle passing from close to the upper border of the medial surface of the thyroid cartilage to the muscular process of the arytenoid cartilage, lying lateral to the fibres of thyroarytenoideus and thyroepiglotticus. Zemlin *et al.* (1988) dissected 15 larynges finding only six superior thyroarytenoid muscles and suggested that its function was to relax the vocal folds, whilst assisting in their medial compression.

It was possible to identify a superior throarytenoid muscle bilaterally in 22 coronal sections of non-humanoid larynges with ventricles (Figure 4.56). Specimens examined included the lemurs, saki monkey (*Pithecia pithecia*), lar gibbon (*Hylobates lar*), chimpanzee and prairie dog (*Cynomys marmota*). Since the number of specimens for each individual species sectioned coronally was limited, any interpretation of the findings must be reserved. It is most probable that this muscle is vestigial and of little functional importance. Kotby *et al.* (1991) also identified a ventricularis muscle within the substance of the ventricular fold in the paraventricular area. They describe it as possessing few fibres, extending along the ventricular wall from the lateral aspect of the arytenoid to the side of the epiglottis. It was identified in 95% of their sections although varying in size and the level at which it was seen. Close to the vocal process of the arytenoid cartilage its fibres deviated laterally to join the most medial fibres of the thyroarytenoid muscle at its insertion into the arytenoid cartilage. This muscle could not be distinguished from the closely related thyroarytenoid muscle in any human or non-humanoid specimen in the collection made by the author, and it should be considered at best rudimentary.

Cricothyroid

In the sixteenth century Andreas Vesalius described the cricothyroid as two separate muscles (Vesalius, 1545). The terms cricothyreoideus rectus and cricothyreoideus obliquus were introduced by Jacob Henle in 1873. However, the notion of two muscles was soon replaced by the concept of a single cricothyroid composed of rectus and oblique bellies (Zaretsky & Sanders,

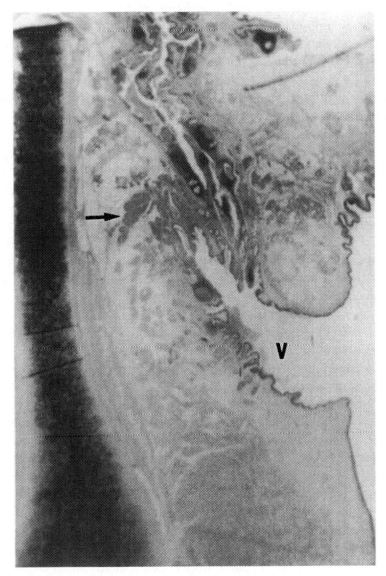

Figure 4.56. Superior thyroarytenoideus muscle (arrowed) in a coronal section of a brown lemur larynx (*Petterus fulvus*). Ventricle (V)

1992). The rectus extends from the anterior ring of the cricoid close to the midline to the inferior border of the thyroid cartilage, with the inferior tubercle as the lateral margin. The oblique fibres originate more laterally on the cricoid edge inserting into the inferior thyroid margin and extending to the inferior horn. This is the classic description of the morphology of this muscle although variations are well recognized. Mayet & Muendnich (1958) described a third part, the pars interna, which extended from the upper cricoid edge to the inner aspect of the thyroid cartilage. In their dissection of eight male mongrel dogs, Zaretsky & Sanders (1992) looked specifically at origins, insertions, fibre orientation and angles of the bellies of the cricothyroid muscle. The origin of the rectus was the anterior border of the cricoid arch 1 mm above the inferior border of the cartilage. The muscle was inserted into the anterior and inner surfaces of the thyroid lamina approximately 10 mm lateral to the midline at the region of the inferior thyroid tubercle. Oblique fibres originated from the inferior border of the cricoid and fanned out to insert on the anterior and inner surfaces of the thyroid cartilage. The oblique line was the lateral margin and unlike the rectus belly appeared to be composed of multiple fascicles. Deep dissection revealed a bundle of muscle fibres with a horizontal orientation. These came from the ventral surface of the cricoid to insert on the inner and medial aspect of the inferior cornu and were present in all dissections and on histological section. Although small and located deep to the principal bellies, this horizontal component was anatomically defined by distinct fascial barriers. They concluded that the varying orientation of each belly conveyed a different mechanical advantage in traction on cricoid and thyroid cartilage.

It is accepted that the cricothyroid plays a major role during phonation and respiration, although absent in the marsupials where cricoid and thyroid remain fused. Arnold (1961) gives average dimensions for this muscle in humans of 15 mm × 10 mm. Whether these dimensions are related to laryngeal size, however, is unknown and to measure the cricothyroid in relation to the cricothyroid complex, dissections were carried out on 63 sagittaly divided non-humanoid larynges. This also provided an opportunity to search for the 'third belly' of the cricothyroid muscle (Figure 4.57). Scaled photographs of formalin-fixed specimens were taken prior to measurement and dissection of the cricothyroid muscle. Following transference of the photographs to the screen of a Macintosh PC and using MacDraw II software, the area of the muscle was calculated. All muscle was removed from cricoid and thyroid cartilages and this area calculated (Appendix 2). Although a significant relationship between body weight and glottic size appears to exist for most mammals with the possible exception of deer, it was unlikely such positive allometry would be found for the cricothyroid muscle (Harrison & Denny, 1985). Regression analysis at the 95% confidence level suggests close grouping at body weights of less than 200 kg, but the correlation coefficient is only 0.504 and this cannot be considered as a significant relationship (Figure 4.58).

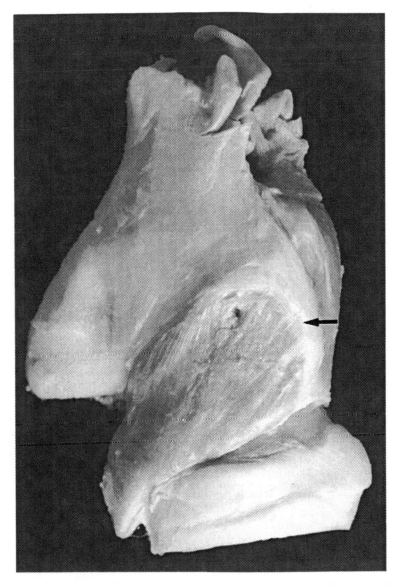

Figure 4.57. Larynx of an adult onager (*Equus hemionus*) cleared of strap muscles to show the cricothyroid muscle (arrowed).

Third belly of the cricothyroid

Dissection of the smaller species is technically difficult, which may explain why the the third belly of the cricothyroid muscle was found rarely amongst them. In all members of the Canidae – coyote (*Canis latrans*), wolf (*C. lupus*), golden jackal

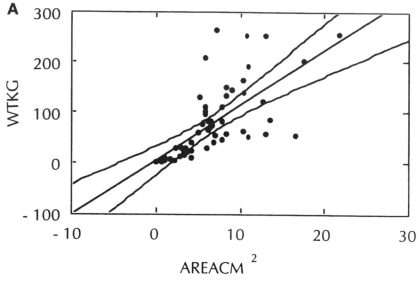

$$WTKG = 5.570 + 10.856 \times AREACM^2$$

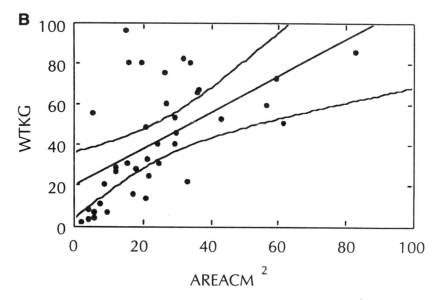

$$WTKG = 19.585 + 0.906 \times AREACM^2$$

Figure 4.58. Regression analysis plots at the 95% confidence levels, (A) for body weight > 100 kg against area of cricoid and thyroid cartilages, (B) for body weight < 100 kg against area of cricoid and thyroid cartilages.

(*C. aureus*) and Cape hunting dog (*Lycaon pictus*) – the muscle was clearly demonstrated. In addition it was found in a single specimen of an American black bear (*Ursus americanus*), Nilgai (*Boselaphus tragocamelus*), pigmy hippo (*Choeropsis liberiensis*), giant panda and chimpanzee. The latter were all adult specimens showing a clear demarcation between the three bellies. In other specimens differentiation proved difficult and it is therefore possible that this variation is present in many species. Whether it plays a separate role in cricothyroid function remains in doubt. (Zaretsky & Sanders, 1992).

These measurements were accumulated from a wide range of species but do not reflect the variations which are to be found in all anatomical morphological data. All specimens were taken from adults but sexual differentiation was impossible because of the restricted numbers involved. Notwithstanding the limitations, the data do present interesting information about the considerable variation in the size of the muscle mass in relation to dead body weight. The measurements had to be two dimensional and varied from 0.8 cm^2 in the tiny rock hyrax (*Heterohyrax brucei*) to 14.1 cm^2 in the yak (*Bos mutus*). When muscle area was related to the percentage of the cricothyroid complex occupied, 18 species fell within the 20% band. The mouflon (*Ovis orientalis*), giant panda and gorilla gave the lowest figures of 11%. Most of the other species fell within the 30% band with the exception of some deer and antelope.

The function of this important muscle in laryngeal function is discussed elsewhere, but as with other anatomical variations detected in this multispecial study, a logical function explanation is not as yet forthcoming. It is likely that they represent normal variations in morphology which would only become apparent with detailed studies of more specimens.

Histochemical characteristics of the laryngeal musculature

It has been observed in many mammalian muscles that the physiological characteristics of motor units can be correlated with histochemical enzyme profiles of their muscle fibres. This provides an indirect means of studying a muscle's activity in both man and other animals (Malmgren & Gacek, 1981). The laryngeal muscles are of necessity, fast acting for airway protection. Hast (1969) demonstrated that like the lateral rectus of the orbit, these muscles show that more than 70% of muscle fibres have multi-motor end plates, bearing a direct relationship to contraction time. However, species differences concerning fibre type composition occur within the laryngeal muscles. Within the cricothyroid, the percentage of fast type II muscle fibres varies from 43% in the sheep to 65% in the rabbit, with humans around 58% (Zaretsky & Sanders,

1992). Slow (type I) and fast (type II) muscle fibres have slightly different myofibrillar adenosine triphosphate which can be differentiated by histological staining. The speed of muscle contraction has also been correlated with the concentration and pH characteristics of actomyosin ATPase activity. Sahagal & Hast (1974) investigated the histochemistry of adult rhesus macaques (*Macaca mulatta*). They found that the thyroarytenoid and lateral cricoarytenoid muscles reacted fastest with contraction times of 14 and 19 msec, both showing high ATPase activity. The cricothyroid and posterior cricoarytenoid were slower with contraction times of 36 and 44 msec, respectively. However, they agreed that the majority of the motor units of the intrinsic laryngeal muscles were composed of fast conducting fatigue resistant fibres. This confirmed the earlier findings of Hast (1968) on the mechanical properties of the thyroarytenoid and cricothyroid muscles in three primates, squirrel monkey (*Saimiri hainanus*), Hyan gibbon (*Hylobates hainanus*) and rhesus macaque. He found no appreciable difference between these species and comparative values were similar to those found for domestic dog but not the cat whose laryngeal muscles were appreciable slower. Comparative studies between human and sheep were made by Zrunek *et al.* (1989), concluding that the sheep larynx showed significantly lower percentages of type I fibres in all except the cricothyroid muscle which they felt was ideally suited for sustained activity.

Whilst histochemistry is a valuable non-invasive indicator of muscle activity, EMG (electromyography) is a true representation of physiological behaviour. As yet no studies correlating electrophysiological, histochemical and morphological variations with individual species behaviour patterns have been carried out. Since the primary laryngeal functions of airway protection and respiration are common to all mammals, it is possible that physiological variations in laryngeal musculature are related to sound production.

Extrinsic laryngeal musculature

Despite differences in general morphology and behaviour of each species these muscles serve to fix the larynx and bring about its vertical movement. Fink (1973) views the action of the strap muscles and pharyngeal constrictors as opening and closing the laryngeal airway by a 'bellows mechanism'. This would open the bellows by pulling the hyoid and thyroid apart, and close by pulling the hyoid down or thyroid upwards. The mammalian strap muscles vary greatly in size although having only one point of attachment to the laryngeal skeleton. Opening and closing of the larynx is complex and whether the 'bellows' hypothesis is relevant to non-humanoid quadrupeds remains debatable.

However, there is considerable similarity between the morphology and physiology of the principal muscles attached to the larynx – the thyrohyoid and
sternohyoid. Prenatally, the primitive infrahyoid muscle mass divides into
superficial and deep layers, with each layer then separating into two. The
sternohyoid and omohyoid result from a split of the superficial layer long-
itudinally into medial and lateral parts. The thyrohyoid and sternohyoid muscles
form as a consequence of the deep layer becoming attached to the oblique line of
the thyroid cartilage. Both thyrohyoid and sternohyoid might thus be considered
functionally as a single muscle mass. However, in the dog the sternohyoid
muscle is larger than the thyrohyoid, exerting three times the force (gram
weight) at twitch contraction, and such variations in muscle size are common in
Mammalia (Hast, 1968).

It has been suggested that apart from fixing the larynx the external muscles
could also influence the shape and size of the glottis. Fink (1973) inferred that
laryngeal depressors such as the sternohyoid, by exerting a stretching force in a
downwards direction on the aryepiglottic folds, indirectly influenced the airway.
Whether such theories can be confidentially extrapolated to non-humanoid
mammals is less certain although the external laryngeal muscles undoubtedly do
play some role in laryngeal function.

4.6 Nerve supply

Galen of Pergamaon (AD 129–199) who was the first anatomist to describe the
recurrent laryngeal nerve as a branch of a cranial nerve demonstrated that if both
nerves were cut in the pig, the squeal was lost. Andreas Vesalius (1555) drew the
cranial nerves and their branches, but it was Willis (1621–75) who provided
details of the anatomy of the vagus nerve and the branches which supplied the
laryngeal musculature (Saunders & Malley, 1982).

The fibres innervating motor units within the intrinsic laryngeal muscles
reach them from their parent cell bodies in the nucleus ambiguus in the medulla
oblongata via branches of the ipsilateral recurrent laryngeal nerve. The cricothyroid muscle is innervated separately through the external branch of the
superior laryngeal nerve whilst the inter-arytenoideus muscle is supplied by
both recurrent laryngeal nerves.

The motor units in the extrinsic muscles are innervated by nerve fibres that
traverse the pharyngeal and cervical plexuses. Finally, it should be noted that the
laryngeal nerves also contain vasomotor and secretomotor fibres of sympathetic
and parasympathetic origin. The former regulate the size of the laryngeal

vasculature whilst the latter control activity within the laryngeal glands (Wyke & Kirchner, 1976).

Morphology of the superior laryngeal nerve

The superior laryngeal nerve leaves the middle of the ganglion nodosum about 36 mm below the margin of the jugular foramen dividing into an external and internal division 21 mm below its origin. Lang *et al.* (1987) found variations in both thickness of the nerve and point of division in their dissections of 44 half heads. Before dividing, the nerve receives filaments from the superior cervical sympathetic chain and pharyngeal plexus. The smaller external branch passes inferiorly on the inferior pharyngeal constrictor and beneath the infrahyoid to supply the cricothyroid muscle. Although invariably sending filaments to the inferior constrictor, small branches may pass to other extrinsic muscles and laryngeal mucosa below the vocal cord.

The larger internal branch passes between middle and inferior constrictor muscles entering the larynx by piercing the thyrohyoid membrane. Occasionally, it may pass through a small foramen in the thyroid cartilage with the superior laryngeal artery. By means of its epiglottic, pharyngeal and communicating branches it supplies the mucous membrane of the larynx, a view substantiated by the findings of Williams (1951). Prior to this many anatomists believed that the superior laryngeal nerve sent motor fibres to the cricothyroid and other muscles, including the inter-arytenoideus. Williams dissected 30 adult normal human larynges obtained at post-mortem and also stimulated the superior laryngeal nerve in two patients during surgery. In each specimen, several filaments from the internal branch of the superior laryngeal nerve were found to enter the inter-arytenoideus muscle. However, histological examination of muscle blocks containing these branches failed to trace them to motor end-plates or muscle spindles. Electrical stimulation produced no detectable movement of the vocal cord, and Williams concluded that branches seen on dissection contained only propioceptive fibres. The descending branch joins the sensory division of the recurrent laryngeal nerve running beneath the mucosa covering crico-arytenoideus posterior to or within the pyriform fossa, to form the Loop of Galen. This provides sensation to the laryngeal mucosa, although in 25% of William's dissections the branches of these tv o nerves were distributed separately and no communicating 'loop' was detectable.

Morphology of the recurrent laryngeal nerve

With elongation of the neck and descent of the heart, the recurrent laryngeal nerves are carried caudally by the lowest persisting aortic arch. The left recurrent laryngeal nerve loops around the ligamentum arteriosum (VI arch) whilst the right-sided nerve, in the absence of the ligamentum arteriosum, passes around the right subclavian artery (IV arch). If this arch is absorbed and the arterial supply to the upper limb is provided by the caudal part of the right dorsal aorta, the nerve passes directly to the larynx (Steinberg et al., 1986).

Classical descriptions of the anatomy of the recurrent laryngeal nerves take little notice of those differences in morphology which may be of considerable clinical significance. Variations in relationship to the thyroid gland and its blood supply are of importance in thyroid surgery, whilst details of the divisions of the anterior motor branch are vital to the development of techniques in laryngeal reinnervation. The right recurrent laryngeal nerve is described as lying at its origin from the vagus in front of the right subclavian artery, hooking below it to ascend in the right tracheo-oesophageal groove. It may cross superficial or deep to the inferior thyroid artery or pass between its branches. The left recurrent laryngeal nerve arises on the left side of the arch of the aorta and looping below it on the left side of the ligamentum arteriosum, ascends behind the aortic arch to the left tracheo-oesophageal groove. It too may have a variety of relationships to the inferior thyroid artery.

Many clinicians and anatomists have investigated the morphology of the recurrent laryngeal nerves confirming that branches supplying the inferior constrictor, cardiac plexus, trachea, cricopharyngeus and oesophagus are of variable size and distribution. The relationship of the nerves to the tracheo-oesophageal groove is said to be 70% within the groove and 30% lateral to the trachea (Crumley, 1982). Steinberg et al. (1986) dissected the necks of 90 fresh, frozen and non-embalmed human cadavers within 48 hours of death, whilst Nguyen et al. (1989) dissected 30 pharyngo-laryngeal specimens from fresh and preserved cadavers. Each studied the anatomy of the recurrent laryngeal nerves with respect to branches destined for the intrinsic laryngeal musculature and the relationship of the nerve trunk to the thyroid gland.

Nguyen et al. (1989) performed dissections of 60 nerves using magnifications of between ten and 40. Extralaryngeal division into an anterior motor branch destined for the intrinsic musculature and a posterior branch representing the inferior part of the ramus communicans was found in 52 recurrent laryngeal nerves (87%). There was no significant sexual variation and the division was situated within 1 cm of the inferior constrictor in 50% of specimens and in the remainder within 2 cm. Steinberg *et al.* (1986) in their dissections of 180 nerves found this division to occur in the upper third of the trunk in 52% of cases but all within 2 cm of the cricothyroid joint. One division they named the cricopharyngeal nerve, this supplied the inferior constrictor and cricopharyngeus muscles. The other, which they named the laryngeal nerve, divides into an anterior motor branch supplying the intrinsic muscles, and a posterior branch sensory to the mucosa inferior to the vocal cord. This anastomoses with the descending branch of the internal laryngeal nerve forming the Loop of Galen.

Nguyen *et al.* continued their dissections of the intralaryngeal anterior branch of the recurrent nerve up to the origin of the interarytenoid nerve. Initially, the anterior branch passed under the inferior edge of the inferior constrictor muscle. One portion of the hypopharyngeal, where the nerve passes through a gap between the inferior cornu of the thyroid cartilage and cricoid, gave a constant innervation to the posterior cricoarytenoid muscle. Superiorly, this nerve also supplied the oblique and transverse arytenoids. The other portion of the nerve innervated the lateral cricoarytenoid ending within the thyroarytenoid muscle. They classified three variations in the nerve supply to the posterior cricoarytenoid muscle. Type I characterized by a single nerve pedicle was the most frequent, occurring in 66.6% of specimens. Other variants with double or triple forking were less common and their data identified both the frequencies, site and length of the pedicles.

The origin of the interarytenoid nerve was found to face the superior half of the inferior cornu in 60% specimens, with the remainder being above the cornu base. This nerve gave collateral branches to the posterior cricoarytenoid muscle and terminated by anastomosing with a branch of the internal division of the superior laryngeal in 3% of cases, and in 5% with the ramus communicans.

The lateral cricoarytenoid nerve was found to have three possible pedicles, all arising from the anterior branch of the recurrent laryngeal nerve. A single pedicle was found in 70% of dissections and bipedicle innervation in 25%. The nerve to the thyroarytenoid muscle rested on the surface of the lateral cricoarytenoid, which in men was 13 mm long and in women 11 mm. Four possible forms of termination were seen but are probably of little clinical significance. Nguyen *et al.* concluded that for selective laryngeal reinnervation it was essential to isolate all pedicles destined for the posterior cricoarytenoid muscle (i.e. all abductor pedicles). In 60% of cases this should be possible following resection of the inferior cornu of the thyroid cartilage.

Relationship of the recurrent laryngeal nerve and its branches to
the thyroid gland and inferior thyroid artery

The incidence of permanent damage to the recurrent laryngeal nerve following thyroid surgery is reported as being between 0.3 and 9%, depending in part on the surgeon's experience and nature of the underlying pathology (Martensson & Terins, 1985). Exposure and protection of the nerve limits accidental injury and knowledge of possible anatomical variations facilitates safer surgery.

The recurrent laryngeal nerve was dissected by Salama & McGrath (1992) in 84 embalmed cadavers to determine the interrelationships between the nerve, posterior fascial attachment of the thyroid gland and the inferior thyroid artery. They found that the fibrous capsule of the thyroid gland divided into two layers posteriorly. One passed posterior to the oesophagus and the other formed the posterior fascial attachment of the gland. The major part of this layer passed between the posteromedial surface of the gland and the anterolateral surface of the cricoid cartilage at the level of the thyroid isthmus, forming horizontal bands. In each dissection the bands blended inferiorly with fibrous adhesions between the gland and adjoining trachea to form a vertical band. The recurrent laryngeal nerve lay close to the two components of the posterior fascial attachment of the gland; dissections showed it ran superiorly, immediately lateral to the vertical band. As the nerve reached the inferior border of the inferior constrictor muscle it came close to the horizontal band and in 68% of dissected nerves it was firmly attached to one or both components of the posterior fascial attachment. In 32% of nerves which were not adherent, the terminal part of the recurrent laryngeal nerve ran lateral to the vertical band and then adjacent to the posterior edge of the horizontal band.

Berry (1888) described a strong band of fascia between the thyroid gland and cricoid cartilage which he called the suspensory ligament. He described the recurrent laryngeal nerve as running on the lateral and posterior aspects of this ligament, although Reeve *et al.* (1969) reported that identification of this structure was unsatisfactory in eight of 157 thyroidectomies. The vertical part of the posterior fascial attachments corresponds to the 'Suspensory Ligament of Berry', and was absent in only three of the specimens dissected. When present the recurrent laryngeal nerve is situated laterally, whilst following the posterior edge of the horizontal band, the cricoid cartilage leads to the nerve.

The inferior thyroid artery in the dissections of Salama & McGrath (1992) and Steinberg *et al.* (1986) was found to have a close but variable relationship to the recurrent laryngeal nerve and its terminal branches. It usually consisted of more than one branch before entering the thyroid gland, and these passed from anterior or posterior or between the branches of the nerve. The least common finding was the nerve passing anteriorly to the artery and the closest relationship between nerve and artery occurred within 2 cm inferior to the cricoid attachment of the horizontal band in 86% of all dissected specimens. In Salama and McGrath's series, the nerve was anterior to the artery in 20% of cases, between arterial branches in 36% and posterior to the artery in 44%. They concluded that there was no constant anatomical relationship between these two structures.

Sensory innervation

The sensory innervation of the laryngeal mucosa is involved in perception of pain, mechanical pressure and chemosensation. Autonomic nerve fibres innervate laryngeal glands and blood vessels and are involved in exocrine secretion and blood flow. Immunocytochemical and physiological studies have demonstrated regulatory peptides which may play a role in non-cholinergic and non-adrenergic systems. Adzaku & Wyke (1979) studied the structure and distribution of sensory nerve endings in the laryngeal mucosa of 27 adult cats by microdissection and neurohistological techniques. The internal branch of the superior laryngeal nerve was found to supply mucosa caudal to the vocal cords, whilst the external branch of this nerve contributed to innervation of subglottic mucosa. After penetrating the cricothyroid muscle, it gave off filaments which passed through the cricothyroid membrane to supply the anterior region of the subglottis. Terminal branches of the recurrent laryngeal nerve supply the remainder of the subglottic mucosa but in the cat it is only to the antero-inferior division, which leaves the parent trunk at the apex of the triangular space between lateral cricoarytenoid and thyroarytenoid muscles.

The mucous membrane lining the larynx contains a mixture of corpuscular and non-corpuscular nerve endings although the density shows considerable variation in disparate areas. It seems probable that differing forms of corpuscular receptors operate as mechanoreceptors and have varying functions; the plexiform and free nerve endings represent a nociceptive or chemosensitive receptor system particularly in the subglottic region. The external laryngeal nerve contains not only motor fibres but afferent fibres from receptors located within the capsules of the cricothyroid joints (Wyke & Kirchner, 1976) and further fibres from receptors within the anterior subglottic region.

The innervation of the larynx by several kinds of neuropeptides in addition to the classical neurotransmitters has been shown by Hisa et al. (1992) in the dog and Hauser-Kronberger et al. (1993) in the human larynx. The latter studied the peptide innervation of the human larynx and recurrent laryngeal nerves using immunocytochemical and radioimmunological methods in tissue taken from 30 patients undergoing laryngectomy. They found that all the regulatory peptides commonly found in the nose and tracheo-bronchial system were present in the human larynx. Although the precise origin of these peptidergic nerve fibres was uncertain, they suggested that most came from intrinsic local ganglia located along peripheral branches of the superior and inferior laryngeal nerves.

Hisa *et al.* (1992) used immunohistochemistry to investigate the distribution of calcitonin gene-related peptide nerve fibres (CGRP) in the laryngeal mucosa, glands and muscles of the dog. They found the amount to vary within the laryngeal mucosa with the greatest density in the epiglottis, subglottis and vocal process of the arytenoid. Both epiglottis and arytenoid are rich in taste buds in the dog and the arytenoid plays an important role in mechanoreception of vocal cord movement. They suggest that CGRP may play an important role in the sensory system of the larynx. The physiological and pathophysiological significance of regulatory peptides in the larynx is far from clear. It seems however, that the complexity of differing laryngeal functions, fine sensations and regulations, are based on a complex system of neurotransmitters and neuromodulators. Clinical disorders such as asthma show, from immunocytochemical and functional studies, that non-cholinergic and non-adrenergic systems are involved using regulatory peptides.

Laryngeal nerve ganglia

The existence of ganglions in the human and canine larynx were reported in the early 1900s, but their precise localization was not discovered until the 1960s, (Yoshikazu *et al.*, 1993). Ganglia were found in the ventricle and paraglottic space, close to branches of the internal laryngeal nerve, and between the cricoid and thyroid cartilages close to the recurrent laryngeal nerve (Figure 4.59). Further ganglia were identified in the laryngeal framework of premature and neonatal babies by Ramaswamy & Kulasekaran (1974). They searched for ganglia on the internal laryngeal nerve in human, canine, feline, rabbit, bat, cow, goat, pig and monkey larynges, identifying a pair of nonchromaffinic ganglia (0.1–0.3 mm in diameter) in the ventricular fold related to the internal branch of the superior laryngeal nerve in human and some pig larynges. Kleinsasser (1964) found a ganglion (0.3–0.4 mm in diameter) related to a branch of the recurrent laryngeal nerve between the caudal end of the cricoid cartilage and inferior horn of the thyroid cartilage. A small ganglion has also been found on both sides of the epiglottis, embedded between fibres of the internal laryngeal nerve supplying the aryepiglottic fold and lower epiglottis. Ganglia have been identified along the course of the internal, recurrent and ramus communicans in canine, rat and cat larynges and are probably present in all mammals (Dahlqvist

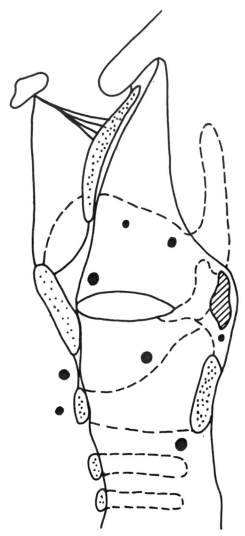

Figure 4.59. Schematic representation of some of the sites within the larynx where glomus tissue (shown as black dots) has been identified.

& Forsgren, 1989) (Figures 4.60 and 4.61). Research has shown that intra-laryngeal ganglionic cells morphologically resemble the carotid body and have neural connections with sympathetic and sensory systems. However, each ganglionic cell group contains a mixture of sensory, sympathetic and parasym-pathetic cells; the functional significance of ganglia within the larynx is yet to be established.

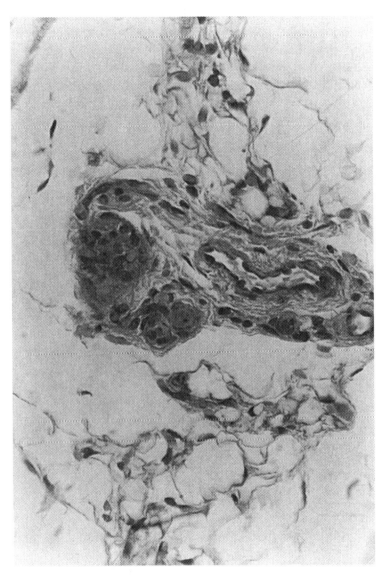

Figure 4.60. Small paraganglia in association with a blood vessel in the larynx of orang-utan (*Pongo pygmaeus*).

Comparative morphology

Although modified to suit individual life-styles the fundamental functions of respiration, phonation and protection are common to all mammalian larynges. Differences can be explained succinctly on anatomical and physiological grounds; this particularly applies to variations in laryngeal nerve supply. Although the superior and recurrent laryngeal nerves carry motor supply to all mammalian intrinsic laryngeal muscles, their routing, branching and anastomosis are variable. Bowden & Schuer (1961) studied the gross pattern of innervation in twenty-six species of eutherian mammals by dissection of

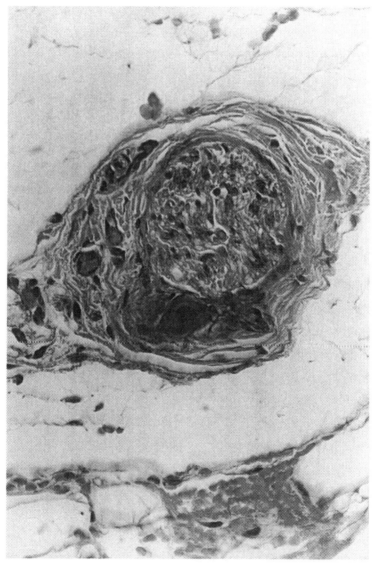

Figure 4.61. Paraganglia in association with nerve fibre in the larynx of orang-utan (*Pongo pygmaeus*).

formalin fixed specimens. Particular attention was paid to the presence of communications between superior and recurrent laryngeal nerves. Where this was present, confirmation of nerve fibres was carried out since in preserved tissue accurate identification was not always possible. Because of the difficulty of obtaining large numbers of less common animals their data was frequently based on single specimens and as with the human larynx, interspecial variations are common. Fourteen human larynges were included in their 63 specimens and in two of these no communication between the internal division of the superior nerve and recurrent laryngeal was found. On the basis of their dissections four different patterns were distinguished (Figure 4.62):

(A) The supply was by the superior and recurrent laryngeal nerves with no connection between them. This was found in six out of nine primate species, and two Insectivora, one Chiroptera and two Lagomorpha.
(B) The internal branch of the superior laryngeal nerve was joined to the recurrent laryngeal by a communication, the ramus communicans of 'Loop of Galen'. This was found in most of the human larynges, two out of the five carnivores, one primate and one artiodactyl.
(C) The internal and recurrent laryngeal nerves were joined by the ramus communicans and a median nerve originating from the superior laryngeal nerve or pharyngeal plexus passed into the superior constrictor muscle. This supply was found in all the artiodactyles except the pig.
(D) This pattern of supply had both internal and recurrent laryngeal nerves and in addition a pararecurrent nerve running parallel with the recurrent and joining the internal branch of the superior laryngeal. This was found in three of the five carnivores and one primate, the marmoset.

Table 4.6 shows that in 21 of the 26 species no more than two specimens of an individual species was available for study. Whilst this in no way detracts from the importance of this research (Bowden & Schuer, 1961), it emphasizes the need for accumulation of more data to confirm the regularity with which a particular supply pattern occurs in a specific species. Variations appear to occur between species belonging to the same order and within individual species. Although there is a fundamental belief that in the development of individual species structure is related to function, there is no apparent functional value for those mammals possessing the more complex pattern of nerve supply. Correlation within zoological orders is unhelpful, since this is based for convenience on morphological factors and is unrelated to behaviour patterns or life-style.

Table 4.7 contains data from dissections carried out on some of the mammals listed in Appendix 1. Selection was restricted because of specimens removed for

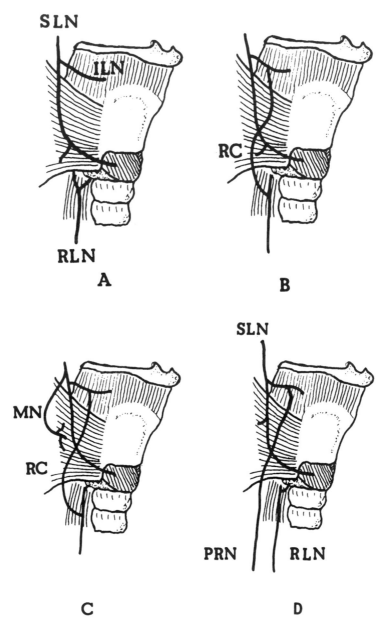

Figure 4.62. Variations in the nerve supply to the larynx in some Eutherian mammals. Key: SLN – superior laryngeal nerve; ILN – inferior laryngeal nerve; RLN – recurrent laryngeal nerve; RC – ramus communicans; MN – median nerve; PRN – pararecurrent laryngeal nerve; A–D refers to the four different patterns identified by Bowden & Scheur (1961). (After Bowden & Scheur, 1961.)

Table 4.6. *Four types of nerve communication identified in eutherian mammals*

Order	Genera	No. of specimens	Type of communication			
			A	B	C	D
Insectivora	*Talpa*	2	2	–	–	–
	Erinaceus	1	1	–	–	–
Chiroptera	*Pteropus*	1	1	–	–	–
	Nyctalus	1	1	–	–	–
Lagomorpha	*Lepus (europacus)*	1	1	–	–	–
	Lepus (cuniculus)	7	7	–	–	–
Primates	*Homo*	14	2	12	–	–
	Hylobates	2	2	–	–	–
	Papio	2	–	2	–	–
	Mandrillus	1	1	–	–	–
	Cerocebus	1	1	–	–	–
	Circopethecus	1	1	–	–	–
	Macacus	1	1	–	–	–
	Saimiri	1	1	–	–	–
	Hapale	2	1	–	–	1
Carnivora	*Felis (domestica)*	5	–	–	–	5
	Felis (serval)	1	–	1	–	–
	Canis (familiaris)	6	–	6	–	–
	Lutra	1	–	–	–	1
	Phoca	2	–	–	–	2
Artiodactyla	*Sus (scrofa)*	3	–	3	–	–
	Bos (taurus)	2	–	–	2	–
Ovis	*Ovis (aries)*	2	–	–	2	–
	Cephalophus	1	–	–	1	–
	Lama	1	–	–	1	–
	Cervus	1	–	–	1	–
Total		63	23	24	7	9

Source: Bowden & Scheur (1961).
A–D: four different patterns of nerve communication as distinguished by Bowden & Scheur (1961).

sectioning or museum preparation, small size, unavailability of enough trachea to identify the recurrent laryngeal nerve or because of difficulties with preservation techniques that inhibited dissection. Histological confirmation of suspected median, recurrent or communicating nerves was carried out in all cases to confirm the presence of nerve tissue; this proved to be negative in eight instances

Table 4.7. *Type of nerve communication identified in selected mammals**

Order	Genera	No. of specimens	Type of communication			
			A	B	C	D
Marsupial	*Macropus* (Bennett's wallaby)	4	4	–	–	–
	(grey kangaroo)	3	2	–	1	–
Insectivora	*Talpa* (European mole)	2	2	–	–	–
	Sorex (common shrew)	2	2	–	–	–
Lagomorpha	*Lepus* (hare)	2	2	–	–	–
	Oryctolagus (rabbit)	6	5	1	–	–
Primate	*Homo*	10	3	7	–	–
	Hylobates (gibbon)	3	3	–	–	–
	Pongo (orang utan)	2	1	1	–	–
	Gorilla	1	1	–	–	–
	Pan (chimpanzee)	3	2	1	–	–
	Cerocebus (mangabey)	2	2	–	–	–
	Theropithecus (baboon)	1	1	–	–	–
	Macaca (Barbary ape)	1	1	–	–	–
Carnivora	*Felis* (domestic cat)	6	2	–	–	4
	(puma)	1	–	–	1	–
	Panthera (lion)	2	–	2	–	–
	(leopard)	2	2	–	–	–
	(tiger)	2	2	–	–	–
	Canis (dog)	6	2	4	–	–
	(wolf)	4	–	4	–	–
	Lycaon (hunting dog)	4	2	2	–	–
	Acinonyx (cheetah)	3	3	–	–	–
Perrisodactyla	*Equus* (common zebra)	3	–	–	3	–
	(onager)	1	–	–	1	–
Artiodactyla	*Sue* (wild boar)	3	2	1	–	–
	Ovis (mouflon)	2	–	–	2	–
	Cervus (Timor deer)	2	1	–	1	–
	Tayassu (collared peccary)	3	2	1	–	–
	Oryx (scimitar oryx)	1	–	–	1	–
	Giraffe	3	–	–	–	3
Total		90	49	24	10	7

*Data from dissections of selected animals, details in Appendix 1.
A–D: four different patterns of nerve communcations as distinguished by Bowden & Scheur (1961).

because of the technical difficulty of dissecting specimens preserved in formalin for long periods.

The differences between the data recorded in Tables 4.6 and 4.7 are expected in view of the absence of any logical explanation for the four patterns of innervation found by Bowden and Schuer (1961). In a series of 14 dogs, Lemere (1932) described a connection between the internal and pararecurrent nerve. Vogel (1952) dissecting the same species found no pararecurrent nerve but described a ramus communicans in all his specimens, a finding confirmed by Bowden and Schuer (1961) and from the author's own dissections. Similar variations in incidence can be expected as more specimens and a wider selection are examined. A variety of innervation patterns exist within the mammalian larynx and although possibly a feature of individual species, is not directly related to laryngeal function. However, the possibility of the neural fibre content of these branches and communications playing a role in the 'fine tuning' of the voice is considered later in section 5.3 on vocalization.

The recurrent laryngeal nerve in the horse

The domestic horse is the product of artificial selection and this process of captive propagation has led to the formation of recognized and recognizable breeds. Most of these were developed in the first instance by a period of intensive inbreeding aimed at establishing characteristics thought to be desirable (Cook & Kirk, 1991). Unfortunately this policy also tends to propagate undesirable characteristics such as recessive diseases, and the restricted genetic pool does not in the case of the thoroughbred lead to increased racing performance. The rationale behind this is discussed further in the section 5.2 on locomotion.

The modern thoroughbred population is descended from a small number of stallions imported from North Africa and the Middle East into England in the 1600s. The most prominent stallions which appear in the genetic picture of a large proportion of modern thoroughbreds are the Godolphin Arabian (b. 1725), the Darley Arabian (b. *c.* 1688) and the Beverley Turk (b. 1690). A fourth horse, the Curwen Bay Barb born about 1699, should also be added to this select list since its genetic legacy to modern horses is also considerable. Between them they have contributed 32% of the genes found in the modern generation of thoroughbreds (Cunningham, 1991). It appears therefore that today's thoroughbred has 'lost' at least 99.99% of its ancestors, and due to the combined effects of deliberate inbreeding and unavoidable genetic drift there has been a further loss of ancestral genes, even from this original and limited collection of alleles.

Idiopathic paresis of the recurrent laryngeal nerve has been a recognizable condition in horses for at least 500 years. Irrespective of language, the colloquial name for the condition has been based upon the noise produced at exercise, 'roaring' or 'whistling'. It was in the nineteenth century that a French veterinarian, Professor Dupuy, discovered that the familiar sound of roaring could be produced experimentally by cutting one recurrent laryngeal nerve. The credit for recording the first accurate account of the condition belongs to a British veterinarian William Percival. In 1824 he described wasting of muscles on the left side of the larynx, although in the sixteenth century 'heavy breathing in horses' had been said to be an inherited condition. It is now clear from the researches of Cook (1976) and others that recurrent laryngeal nerve neuropathy (RLN) is present to some degree in most thoroughbreds and is an inherited, incurable and progressive disease (Figure 4.63). This usually occurs on the left side but may affect the right nerve initially or present bilaterally. Complete paralysis is present in fewer than 7% of racehorses, whereas 93% may have some degree of unilateral partial paresis (Cook, 1976). Since this will not produce the dramatic sound of 'roaring', many horses are not diagnosed but are considered as having a poor racing performance. The need for a fully functional glottis in running is considered elsewhere but modern thoroughbreds, even with a normal larynx, are probably already at the limit of their racing performance. Any reduction in laryngotracheal airflow will have a dramatic effect on both speed and endurance (Derksen et al., 1986). Duncan et al. (1974) found some evidence of muscle atrophy in 90% of adult horse larynges at post-mortem examination. Early clinical diagnosis is possible by 'palpation' of the larynx, although in the absence of an effective surgical cure this may only be of limited economic value.

Venticulectomy (the Williams or Hobday operation) developed in Germany in the late nineteenth century was rapidly abandoned by its inventor but popularized in the USA and subsequently in England by Professor Hobday. Essentially an avulsion of the ventricle on the side of the paresis, its purpose was to produce vocal cord lateralization from subsequent scarring (Figure 4.64). Recent surgical procedures developed for treatment of human laryngeal disease have now been applied to the equine larynx. Despite the ease of operating on this large organ, long-term success is limited, for the racehorse requires fully abducted vocal cords to maximize glottic airflow. 'Roaring' can be stopped but performance remains restricted.

Variations in fibre size within the laryngeal nerves
Galen suggested that the brain governed movement and sensation by secreting a psychic 'pneuma' that travelled through imperceptible channels within nerves.

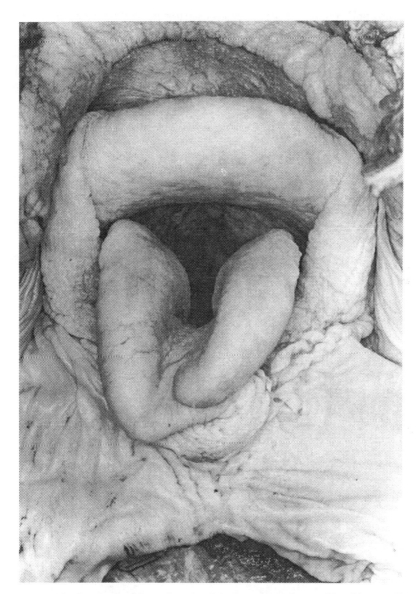

Figure 4.63. Paralysis of the right side of the larynx in a thoroughbred horse. The right arytenoid has fallen laterally following paresis of the vocal cord.

Even within the eighteenth century, the brain was considered a glandular organ with nerves carrying brain secretions to the periphery. Although physiologically incorrect, substances that originate in the body of a nerve cell are delivered to the axon's terminals by passage through the lumen of a fibre. Fibre size offers a

convenient and practical means for comparing nerve fibres and external fibre
diameter, a term which includes the axon and myelin sheath, that can be used to
classify and compare myelinated fibres. The insulating properties of myelin are
essential for rapid conduction of nerve signals and the velocity of conduction of a
nerve sheathed in myelin approximately increases with fibre diameter. This is in

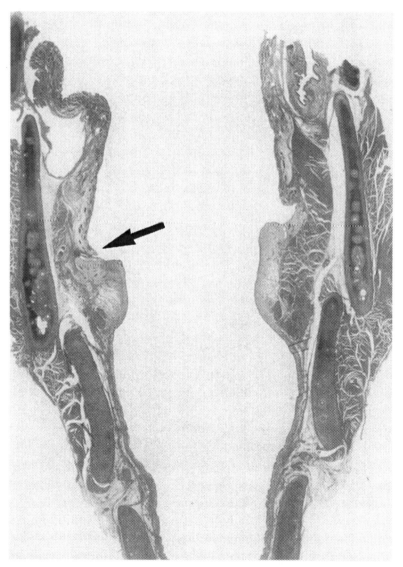

Figure 4.64. Coronal section of the larynx of a horse following a right sided 'Hobday'
operation. The ventricular space has been obliterated by fibrous tissue (arrowed).

contrast to non-myelinated fibres where the relationship is to the square root of
the diameter. The electrophysiological excitability, conduction velocity, action
potentials and biochemical parameters of myelinated nerves are related to fibre
diameter. In view of the disparity in size and pattern of laryngeal innervation in
human and other mammals, estimations of nerve fibre size frequency have been
carried out in expectation of providing greater understanding of the mechanism
of laryngeal function.

Methods of measuring nerve fibres   Previously, attempts at evaluating fibre-size
distribution in laryngeal nerves have been restricted by inadequate techniques,
most of which were time-consuming and inaccurate. Methods used for measuring transversely sectioned nerve fibres have ranged from simple point counting
techniques and the use of circles of known diameters, to semi-automatic and
automatic measuring devices.

Not all fibres of a transversely sectioned nerve bundle are cut transversely,
since many pursue a tortuous course within a given bundle (Figure 4.65).
Features which suggest that a fibre has been cut obliquely include an elliptical
shape, the myelin sheath being thicker at the ends of the long rather than the

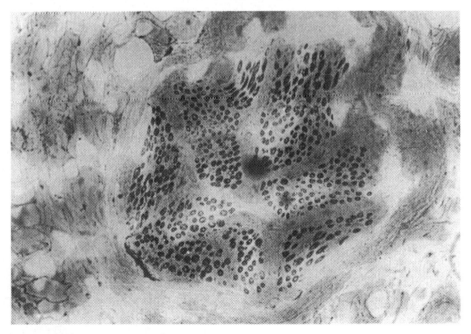

Figure 4.65. Transverse section through a human recurrent laryngeal nerve 2 cm
below lower border of the cricoid cartilage.

short diameter and elliptical neurotubules. Fraher (1980) estimated that if the plane of section is 10° on either side of the transverse plane, the overestimate is just under 1% for circumference and 1.85% for estimates of area. These errors can never be completely excluded, although automatic systems allow elimination of gross obliquity and quantification of a variety of parameters, such as circumference, area, diameter, etc.

Many workers have estimated fibre sizes by measuring all myelinated fibres in a complete transverse section through each nerve trunk. The effort involved in this approach depends upon total fibre numbers and can be considerable. Mayhew & Sharma (1984) give the number of myelinated fibres in the sciatic nerve of the mouse as 2300 and 107,000 in the optic nerve of the rat.

The advantages and limitations of machine measurements have been subjected to statistical evaluation by Mayhew (1983) and others, concluding that in nerves with a large number of fibres, sampling is more accurate than attempts at measuring every fibre. In terms of time, reproducibility and versatility Fraher (1980) concluded that semi-automatic optic analysing systems were the most efficient with relatively unimportant systematic errors, providing data comparisons were made using the same method. It is these variations in sampling techniques and methodology that make historical data comparisons of laryngeal nerves difficult to evaluate.

Human laryngeal nerves   Most measurements have been made on small numbers of nerves obtained at post-mortem. Following fixation in osmic acid, which stains the myelin sheath black, sections are photographed and then magnified. This allows comparisons with known diameters (Scheur, 1964), but is a slow and laborious technique and has been replaced by the image analyser linked to a computer with a maths coprocessor. This allows rapid and reproducible measurements of a variety of fibre parameters (namely, area, diameter, circumference).

Errors are present in all methods of nerve fibre measurement with many publications failing to record the level at which nerves were sectioned and comparisons being made with data obtained by different methods of measurement or with low levels of reproducibility (Fraher, 1980). This probably accounts for many of the major differences found in published reports. Murtagh & Campbell (1951) reported on fibre counts in four human recurrent laryngeal nerves without giving any of the levels of section. The number of myelinated fibres ranged from 1598 to 2891, the largest being 17 μm in diameter. Over 30% were less than 3 μm and histograms showed bimodality, the peaks lying at 2.8 and 8.0 μm. They considered that this showed some similarity with cat and goat, although the latter has more large fibres. Tomasch & Britton (1955) based their findings on one 32-year-old man killed in an accident. Site of section of both the superior and recurrent laryngeal nerves and specific branches were identified and direct measurements made from microphotographs. The left recurrent was composed of 10 bundles and the histogram showed a major peak at 2 μm, with few fibres greater than 8 μm. The largest diameter was 9 μm. Sections taken close to the anastomosis of the internal branch of the superior laryngeal nerve and recurrent nerve showed two peaks in their fibre count. One at 1–2 μm and the other 6–7 μm. From this Tomasch & Britton (1955) concluded that the intrinsic laryngeal muscles must be supplied via the internal laryngeal nerve. No data for the number of fibres counted was included in this paper.

Scheur (1964) gave the site of section for both internal and recurrent laryngeal nerves in three hemilarynges. Using magnification of × 750 he measured nerve diameter with a Perspex template finding a unimodal fibre-size distribution for the recurrent nerves, with total fibre counts of 1493, 788 and 687. The peak was 10–12 μm with a spread of 2–16 μm. The internal laryngeal nerve was also unimodal peaking at 2–4 μm with a spread of 2–16 μm and total fibre counts of 2012, 3646 and 4668. Since the ramus communicans is variable in its presence he considered it unessential in laryngeal innervation. However, he believed that the sum of the myelinated fibres in internal and recurrent nerves gave a more realistic evaluation of the laryngeal neural complex. When this was calculated for his three cases, figures of 3503, 4434 and 5355 were obtained. High counts in one internal nerve were compensated for by a low count in the recurrent on the same side. This supported the hypothesis of Dilworth (1921) that 'these nerves were a plexus of the vagus represented by a continuous nerve joining the internal and recurrent nerves and that separation from this strand formed the individual nerves of the larynx'.

Harrison (1981) counted the myelinated fibres in the recurrent laryngeal nerves of six human larynges, sections being taken 2 cm from the inferior border of the cricoid cartilage. Measurements were made with an Optomax Image Analyser System III with software carried by a IBM compatible computer. The programme allowed differential counting in two micron steps, each measurement being the mean of 10 counts. Five counts were made on each field and the results computed with production of data and histograms. Fibre distribution for both recurrent nerves was unimodal with a peak between 6–10 μm on the right and 8–16 μm on the left, with a spread of fibre size between 2 and 16 μm. Average count for the right nerve was 1238 and on the left 1247 (Tables 4.8 & 4.9, Figures 4.66 and 4.67). Although the nerve sections were carefully prepared some distortion in circumference was unavoidable. The design of image

Table 4.8. *Numbers of myelinated fibres in each group of humans: right recurrent laryngeal nerve*

	Number of myelinated fibres in each group								
Patient	0–2 μm	2–4	4–6	6–8	8–10	10–12	12–14	14–16	Total
S.P.	12	61	84	331	364	237	149	104	1342
B.L.C.	8	46	102	264	302	193	121	88	1124
W.P.	8	48	84	283	317	209	130	89	1168
R.W.	12	56	80	295	324	211	131	93	1202
L.R.	11	55	86	309	351	224	147	101	1284
T.R.S.	2	60	81	324	341	231	145	122	1306
Total (%)	0.8	4.4	6.9	24.3	26.9	17.6	11	8.1	Average (1238)

Table 4.9. *Numbers of myelinatedd fibres in each group of humans: left recurrent laryngeal nerve*

	Number of myelinated fibres in each group								
Patient	0–2 μm	2–4	4–6	6–8	8–10	10–12	12–14	14–16	Total
S.P.	8	53	95	262	332	234	171	131	1286
B.L.C.	4	48	103	251	286	187	182	146	1207
W.P.	4	39	126	197	328	192	168	138	1192
R.W.	11	47	123	291	254	263	189	116	1294
L.R.	2	34	90	202	312	237	168	128	1173
T.R.S.	8	63	112	279	399	189	138	142	1330
Total (%)	0.5	3.8	8.7	19.8	25.5	17.4	13.6	10.7	Average (1247)

intensifiers, however, minimizes observer error and represents the most effective means of measuring fibre area, projection length, diameter and intercept.

The data in Tables 4.8 and 4.9 indicate a preponderance of large myelinated fibres in the left recurrent laryngeal nerve at this level of section; this may be responsible for increased conduction velocity in the longer of the two recurrent laryngeal nerves.

Comparative studies of fibre-size frequency   During a quantitative study of the cat's vagus nerve, Murray (1957) also counted the number and size of the fibres within the recurrent nerves. Sectioned at the level of the clavicle the nerve

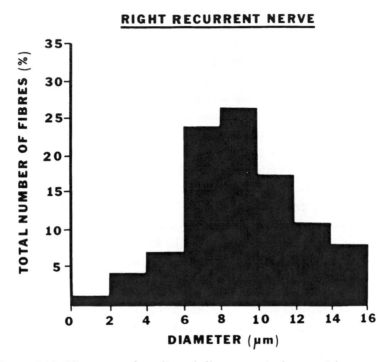

Figure 4.66. Histogram of myelinated fibre count in human right recurrent laryngeal nerve. (From Harrison, 1981.)

comprised two groups of fibres, one with large myelinated fibres the other containing small moderately myelinated fibres. As the nerve ascended between the oesophagus and trachea most of these small fibres were distributed to these structures. Measurements in five nerves gave a unimodal size-frequency peaking between 10 and 12 μm in diameter with a mean of 462 myelinated fibres in the larger bundle. At the level of the clavicle there was a mean of 566 small fibres, showing a unimodal distribution with a peak of 4–6 μm. He concluded that the fibre content of the recurrent laryngeal nerve in the cat was similar to that of the rabbit (Evans & Murray, 1954), with few fibres above 14 μm or below 6 μm in diameter.

Abo-El-Enein (1967) measured the fibre-size distribution in adult cats at a level 1 cm caudal to the cricoid cartilage, and also the internal laryngeal nerve distal to the bifurcation of the superior laryngeal nerve. The total number of myelinated fibres in the four nerves studied was 545, 744, 650 and 962 respectively, with diameters ranging from 2 to 16 μm. The fibre diameter was bimodal with peaks at 2–4 μm and 10–12 μm. He also found that the larger of

164 Nerve supply

the two branches consisted mainly of fibres greater than 6 μm. The total number of fibres within the internal laryngeal nerves were 1234, 751, 1545 and 1389 respectively, diameters ranging from 2 μm to 16 μm showing a bimodal pattern.

Gacek & Lyon (1976) in their experimental investigation of the motor and sensory components of the recurrent laryngeal nerves in 21 cats, measured fibre size using light and electron microscopical sections. Photographs were taken and after enlargement, fibres were counted and diameters measured. Average totals of 565 fibres in the right nerve and 482 in the left compare well with the counts of Murray (1957) and Abo-El-Enein (1967), although Murtagh & Campbell (1951) had found 1026 and 1456 in their three cats. The fibre diameter spectra also compared favourably with other publications. They distinguished three groups of myelinated fibres within the recurrent laryngeal nerve. One contained fibres smaller than 3 μm in diameter, a second group had small numbers of large fibres measuring 10–15 μm and a third had numerous fibres of 4–9 μm. Subsequent experiments confirmed that all the efferent motor fibres were contained in the latter group. They concluded that the small fibre group contained afferent fibres supplying tracheal and oesophageal mucosa

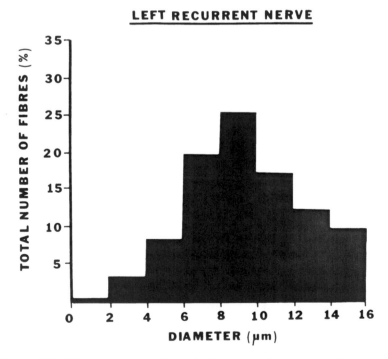

Figure 4.67. Histogram of myelinated fibre count in human left recurrent laryngeal nerve. (From Harrison, 1981.)

whilst the largest fibres measuring 10–15 μm, were probably afferents coming from the superior laryngeal via the ramus communicans. Unmyelinated fibres in the nerve either represented sympathetic fibres from the cervical sympathetic chain or postganglionic parasympathetic fibres from cell bodies at the skull base. These would provide innervation for laryngeal blood vessels and glandular tissue.

Dahlqvist *et al.* (1982) carried out a quantitative study of the recurrent laryngeal nerves in the rat from their origin from the vagus to entry into the larynx. The diameter of myelinated fibres displayed a bimodal pattern with peaks at 2–3 μm and 4.5–7 μm. No difference was found between the number of fibres in the right and left nerves, but wide variations existed in the total number of fibres counted between different nerves. The lowest count was 192 in a right-sided nerve, the highest, 542, was in a left nerve.

Of particular interest is the quantitative measurements reported by Lopez-Plana *et al.* (1988) on sections taken from the distal part of the recurrent laryngeal nerves of 18 mixed-breed horses of varying ages. Total counts of myelinated fibres in the left and right nerves were significantly different at 965 and 1043 respectively. Mean fibre diameters in the left and right nerves (9.53 μm and 10.48 μm respectively) were also significantly different; the distribution was unimodal. Fibre diameters ranged from 2 to 17 μm with 91% having a diameter greater than 5 μm. In view of the suggestion by Cook (1976) and others that idiopathic laryngeal hemiplegia is subclinical in between 20 and 40% of all 'normal' horses, it is possible that reduction in number of myelinated fibres at the distal level of the left recurrent laryngeal nerves is secondary to this condition, as reported by Duncan *et al.* (1974)

It is difficult to draw more than general conclusions from these studies of fibre-size distribution in a such a small number of mammalian species because of variations in measuring techniques and lack of data relating to site of section. All mammals, however, have recurrent laryngeal nerves longer on the left than the right. Differences may vary from 0.8 cm in the rat, 13 cm in dog, 11 cm in humans to over 30 cm in the giraffe (*Giraffa camelopardalis*) (Dahlqvist *et al.*, 1982). Average lengths of the left nerve in man are 43 cm and right 32 cm, although the range extends from 5.7 to 15 cm (Shin & Rabuzzi, 1971). The left recurrent laryngeal nerve of the giraffe is the longest nerve in the animal kingdom being over 2 m long. Other factors being equal, a variation in length of conduction pathway from nucleus ambiguus to intralaryngeal musculature of 30 cm should result in a discernible lag of the left vocal cord behind that on the right; this does not occur in the giraffe or any other mammal.

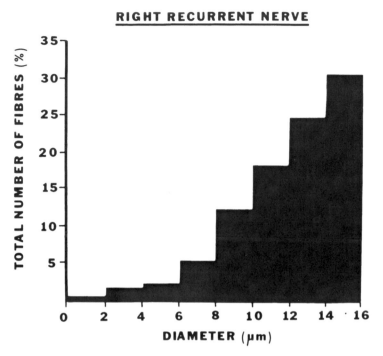

Figure 4.68. Histogram of myelinated fibre count in the right recurrent laryngeal nerve of an adult giraffe. (From Harrison, 1981.)

Conduction in the recurrent laryngeal nerves

A possible explanation for the simultaneous arrival of both nerve impulses at the laryngeal musculature would be faster conduction in the left recurrent laryngeal nerve. Harrison (1981) measured the fibre-size frequency of myelinated fibres in recurrent laryngeal nerves sectioned 12 cm below the lower border of the cricoid cartilage in two giraffes. He found a unimodal distribution with definite preponderance of large fibres in the left nerve. The single peak lay in the 6–12 μm range but 10% of fibres with diameters between 14 and 16 μm were in the left nerve (Tables 4.10 and 4.11, Figures 4.68 and 4.69). This animal has a pararecurrent nerve and in the two available for measurement all fibres were less than 6 μm in diameter (Figure 4.70). This preponderance of large fast conducting fibres in the left nerve may be necessary to counteract any delay resulting from the disparity in length between the two recurrent laryngeal nerves in this animal.

Shin & Rabuzzi (1971) studied conduction velocity in the canine recurrent laryngeal nerves between the vagus at the level of the cricoid cartilage and the

Table 4.10. *Numbers of myelinated fibres in each group of giraffes: right recurrent nerve*

Giraffe	Number of myelinated fibres in each group								Total
	0–2 μm	2–4	4–6	6–8	8–10	10–12	12–14	14–16	
Male 15 yrs	5	12	14	49	116	206	232	310	944
female 13 yrs	9	16	19	60	122	139	241	287	893
									Average
Total (%)	0.9	1.5	1.8	5.9	12.9	18.8	25.7	32.5	(919)

Table 4.11. *Numbers of myelinated fibres in each group of giraffes: left recurrent nerve*

Giraffe	Number of myelinated fibres in each group								Total
	0–2 μm	2–4	4–6	6–8	8–10	10–12	12–14	14–16	
Male 15 yrs	4	11	26	40	85	118	291	314	889
female 13 yrs	2	9	23	58	85	129	310	301	917
									Average
Total (%)	0.4	1.1	2.7	5.4	9.4	13.7	33.3	34.0	(903)

nerve at the 13th tracheal ring. The right nerve averaged 58.0 ± 6.7 msec whilst the left averaged 65.5 ± 2.8 msec. Lengths of nerve studied were 15.5 to 27.5 cm on the right side and 31.5 to 45.0 cm on the left. Despite problems when measuring the nerves they concluded that the conduction velocity on the right side was slower than the left by approximately 7.5 msec.

Measurements of conduction time in the human present problems of access to adequate lengths of recurrent nerve but has been performed by Peytz *et al.* (1965) on 29 patients. The stimulus was applied to the recurrent nerve close to the trachea and vagus prior to the division of the nerve. A conduction time of 2.1 ± 0.05 msec over a conduction distance of 6 cm corresponds to a latency of 2.7 ± 0.02 msec in the facial nerve, where conduction distance is longer. In eight individuals conduction velocity averaged 65 msec and was identical in right and left nerves over distances of 10 to 20 cm.

Despite lack of confirmation of differences in conduction times or velocity between the two recurrent laryngeal nerves, the left nerve is always longer than the right. This must result in a disparity of impulse conduction time unless some compensatory mechanism is present. Faster conduction times in the longer nerve would serve to explain the disparity in length, and there is evidence that myelinated fibre diameters tend to be greater on the left side. Myelinated fibre diameters in peripheral nerves correlate well with impulse conduction velocity because the nodes of Ranvier are more widely spaced in larger fibres. O'Reilly & Fitzgerald (1985) investigated internodal lengths and length/diameter ratios in 1 mm segments from the laryngeal nerves of a 54-year-old man who had died from accidental poisoning. They found no evidence that the left recurrent laryngeal nerve compensated for its greater length (170 mm from point of origin compared to 70 mm in the right nerve) by having longer internodes for fibres of a given diameter. To provide for simultaneous conduction rates in the two nerves they calculated that the internodal lengths in the left nerve would have to be more than twice as long as on the right.

Why then do the recurrent laryngeal nerves not elongate during the sixth and tenth weeks of prenatal development similarly to those in the cauda equina? The answer may lie in the timing of the two events, for the fourth aortic arches descend during the fifth to tenth weeks, whereas the spinal cord ascends during the thirteenth to fourteenth weeks (Barson, 1970). Peripheral nerve myelination begins during weeks 16 to 20, after the recurrent laryngeal nerves have reached

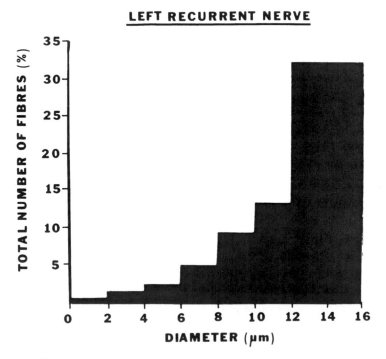

Figure 4.69. Histogram of myelinated fibre count in the left recurrent laryngeal nerve of an adult giraffe. (From Harrison, 1981.)

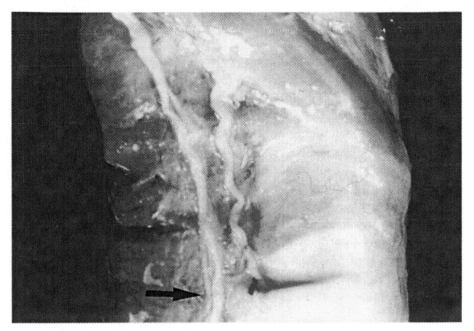

Figure 4.70. Macroscopical dissection of the larynx of a giraffe showing the recurrent laryngeal nerve entering the larynx behind the cricoid cartilage. The pararecurrent laryngeal nerve (arrowed) is passing superiorly to join the internal branch of the superior laryngeal nerve.

their final positions, but during elongation of nerve roots in the cauda equina. At a functional level, impulse signals in the motor fibres to the laryngeal muscles in humans would take only 3 msec longer on the left side in fibres of 10 μm diameter. This delay may not be significant since a further 350 msec are spent building up sufficient subglottic pressure for vocalization. Whether this matters in a largely non-vocal animal such as the giraffe is doubtful.

Considerable interest has been shown and research carried out into the neurological systems controlling the activity of intrinsic and extrinsic laryngeal musculature related to respiration, locomotion and vocalization. Although this is discussed in the chapter devoted to Laryngeal Physiology, there is a lack of knowledge of the comparative morphology of the neuro-anatomy of the laryngeal nerves. This deficiency limits our understanding of the subtle variations of function which are such a noticeable feature of the mammalian larynx and can only be remedied by further comparative studies.

4.7 Vascular supply

The description by Terracol & Guerrier (1951) of the gross arterial supply to the human larynx still remains adequate for most surgical procedures. More recently, development of conservation laryngeal surgery together with an expectation that detailed knowledge of the endolaryngeal blood supply would assist in the understanding of tumour spread, has led to studies on laryngeal microcirculation. Both Andrea (1975) and Pearson (1975) separately published detailed accounts of the microvascular pattern within the human larynx. The former studied 150 cadaveric specimens by microangiography followed by dissection, sectioning or corrosion casting. Pearson examined 20 human hemi-larynges by microdissection following injection of a coloured water-based colloidal suspension of 1 μm radio-opaque (Micropaque) particles. This was made up as a 5% gelatin solution and following infusion and fixation, specimens were cleared in silicone fluid following the method used by Harrison (1971) to study the nerve supply of the maxilla. Eight specimens were serially sectioned at 2 mm following differential staining to determine whether filled vessels were arterioles or capillaries.

Andrea continued his studies, concentrating on the vascular supply of the anterior commissure. In 57 human larynges the microvasculature was examined by micropaque injection; half of the larynges were dissected whilst the remaining specimens were sectioned at 500 μm.

Nomenclature

Although there is not complete agreement on precise terminology, the endolarynx receives blood from both superior and inferior laryngeal arteries as well as from a small posterior–inferior vessel (in 98.25% of 114 cases) which arising from the inferior thyroid, runs under the mucosa of the anterior wall of the laryngopharynx. The superior artery is invariably present as a single vessel (incidence of 94.90%) arising from the superior thyroid artery, entering the larynx through the thyrohyoid membrane. Andrea found that in 5.38% of specimens it passed through a foramen thyrodeum (Figure 4.5 and 4.6) before entering the endolarynx. Although variations in the divisions of the superior laryngeal artery are relatively common, Pearson recognized five major branches

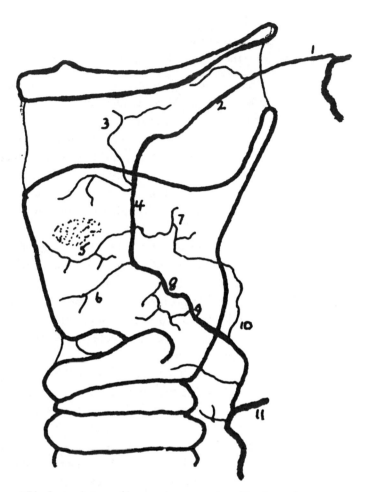

Figure 4.71. Lateral view of human larynx to show blood supply:

1 Superior thyroid a. 2 Superior laryngeal a. 3 Ascending branch of superior laryngeal a. 4 Vental branch. 5 Medial branch to ventricle. 6 Dorsal branch. 7 Anterior division of descending branch. 8 Posterior division. 9 Lateral division of inferior laryngeal a. 10 Medial division. 11 Inferior thyroid a. (After Pearson, 1975).

(See Figure 4.71):

1. Ascending – this passes across the pyriform fossa towards the lateral edge of the epiglottis.
2. Ventral – related to the ventricle and saccule when present.
3. Medial – penetrates the false cord.
4. Dorsa – crosses the floor of the pyriform fossa to the postcricoid region.

5. Descending – continues inferiorly to the thyroarytenoideus muscle where it divides into anterior and posterior divisions. These communicate with ascending branches from the inferior laryngeal artery.

The inferior laryngeal artery shows variations in its branching which are described in detail by Andrea (1975). Most commonly it divides into two: a medial vessel which crosses the posterior cricoarytenoid muscle, joins the dorsal branch of the descending division of the superior laryngeal artery and then enters the larynx via the cricothyroid membrane; and the lateral branch which ascends along the cricothyroid muscle to anastomose with the posterior division of the descending branch of the superior laryngeal artery. Either or both of these branches may be absent, with the lateral division being found in only 36% of specimens (Andrea, 1975).

A vascular arcade, formed by medial branches from both the inferior laryngeal vessels lying superficial to the cricothyroid membrane, sends branches to the subglottic mucosa and then extends superiorly to the anterior commissure. Midline and lateral vessels perforate the membrane to supply the perichondrium of thyroid and cricoid cartilages; other branches reach the cricothyroid muscle. Occasionally the posterior aspect of the thyroid cartilage is perforated by a branch or the whole of the superior laryngeal artery, this occurred in approximately 3% of specimens. (See Figure 4.72.)

Anastomosis between branches of superior and inferior laryngeal arteries form two longitudinal and one transverse arcades, from which branches arise to supply the endolarynx. Since a variety of vascular patterns are found in both human and other mammalian larynges, specific naming of many of these small terminal vessels is not always possible. Despite this lack of a consistent vascular arrangement within the larynx, the supply to some areas appears to be fairly constant. Blood supply to the laryngeal surface of the epiglottis comes from the periphery via the ascending branch of the superior laryngeal artery on the inner surface of the thyrohyoid membrane. This vessel gives small branches to the pre-epiglottic space. An additional branch comes from the lingual artery after supplying the lingual surface of the epiglottis.

The ventral branch of the superior laryngeal artery arises within the paraglottic space giving off a number of small vessels supplying perichondrium, saccule, petiole and finally the pre-epiglottic space. The medial branch supplies all of the false cord including the corniculate and cuneiform cartilages.

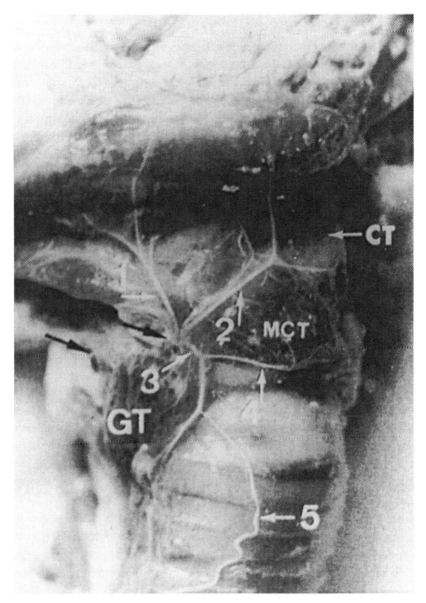

Figure 4.72. Lateral dissection of the human larynx following injection of radio-opaque material showing:

1 muscular branches. 2 branch to cricothyroid. 3 branch of thyroid gland. 4 branch to cricotracheal membrance. 5 arcade over trachea. CT thyroid cartilage. MCT cricothyroid muscle. GT thyroid gland. (After Andrea, 1975.)

The vocal cord and laryngeal mucosa

Because of the clinical importance of inflammation and trauma to the surface of the vocal cords, considerable attention has been paid to their structure and blood supply. The subglottic mucosa receives most of its blood supply from the cricothyroid arcade, vessels penetrating the cricothyroid membrane reach as far superiorly as the inferior surface of the vocal cord. An additional supply comes from the anterior division of the descending branch of the superior laryngeal artery after penetrating the conus elasticus. This is part of the longitudinal arcade and also supplies the true vocal cord with arterioles and capillaries which reach the cordal edge from either end of the membranous vocal fold. Small vessels run parallel to the edge of the cord, whereas larger arteries can be found on both the superior and inferior surfaces of the vocal cord, as is seen in acute laryngitis (Hirano & Sato, 1993). The conus elasticus provides an almost avascular barrier between the subglottic vessels and many small arteries within the cricothyroid and cricoarytenoid musculature. Andrea (1981) has shown by microangiography that a small avascular zone also exists at the origin of the vocal cords in the thyroid cartilage. The periarteriolar pathways from both the vocalis muscles pass inferiorly to the anterior commissure before uniting with each other, confirming that there is no morphological separation between glottis and subglottis at this region, which he terms the 'plane zero'.

Soon after entering the larynx the inferior laryngeal artery sends branches to the posterior cricoarytenoideus muscle and overlying mucosa. The medial division ascends to anastomose with the dorsal branch of the superior laryngeal artery, forming a longitudinal arcade (Figure 4.73). This supplies the inter-arytenoideus muscle and small anastomoses are also found around the cricoarytenoid joints and related muscles. These are so variable that specific naming of individual vessel is unhelpful.

Similar investigations have been carried out on several other species, confirming that the principal blood vessels supplying the mammalian larynx are similar to those of the human. It appears likely that the main divisions are also similar, since they are related to the existence of morphological structures common to all larynges (Figures 4.74 and 4.75).

Pearson (1975) felt that since blood vessels and lymphatics permeate similar tissues within the larynx, they might provide pathways of reduced resistance to the spread of cancer. Certainly, there is evidence that correlation exists between the morphology of intralaryngeal vascularity and the routes taken by endo-laryngeal cancer. However, the frequency with which variations in the presence and routing of specific vessels occur in human and other mammalian larynges,

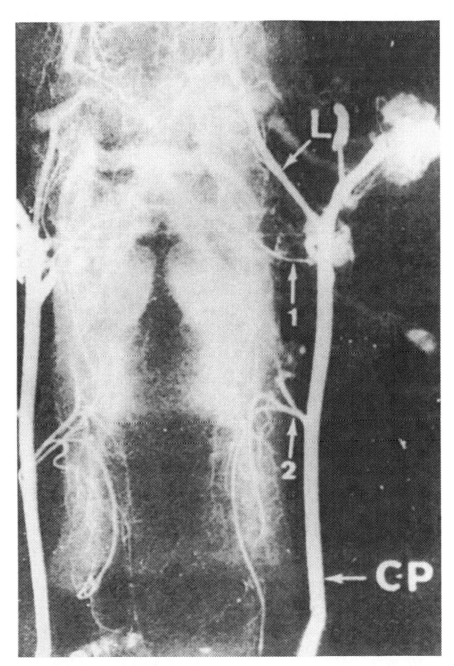

Figure 4.73. Arteriogram of a human larynx showing longitudinal arcades: **1** superior laryngeal artery. **2** inferior thyroid artery. **CP** common carotid artery. **L** lingual artery. (After Andrea, 1975.)

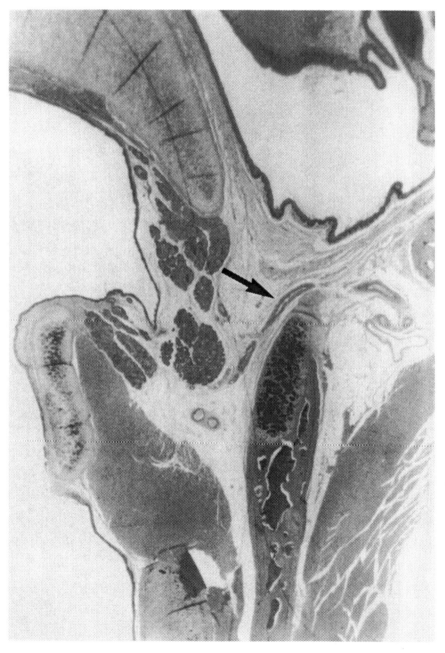

Figure 4.74. Coronal section of the larynx of common wombat (*Vombatus ursinus*) showing ventral branch of superior laryngeal artery in superior region of paraglottic space (arrowed).

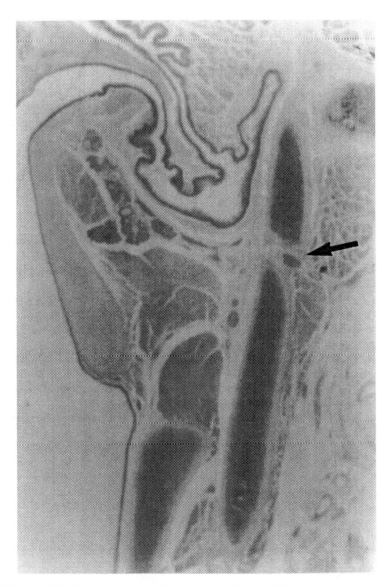

Figure 4.75. Coronal section of the larynx of coypu (*Myocastor coypus*) showing superior laryngeal artery entering larynx through foramen in the thyroid cartilage (arrowed).

suggests that cross circulation may be sufficient for adequate vascularization. Detailed nomenclature of peripheral branches may indicate a restriction of vascular supply which in practice has only limited physiological importance.

References

4.1 Development of the larynx

Born, G. (1883). Die Plattenmodellirmethode. *Archives of Anatomy*, **22**, 584–99.

Goedbloed, J. F. (1960). *De Vroege Ontwikkeling van Let*. Leiden, Middlemoor.

Magriples, V. & Laitman, J. T. (1987). Developmental change in the position of the fetal human larynx. *American Journal of Physical Anthropology*, **72**, 463–72.

Müller, F., O'Rahilly, R. & Tucker, J. A. (1985). The human larynx at the end of the embryonic period proper. *Annals of Otology, Rhinology and Laryngology*, **94**, 607–17.

Negus, V. E. (1924). The mechanism of the larynx. *Lancet*, **207**, 987–93.

O'Rahilly, R. (1973). *Developmental Stages in Human Embryos, Including a Survey of the Carnegie Collection*. Part 1: *Embryos of the First Three Weeks (Stages 1 to 9)*. Washington DC, Carnegie Institution.

O'Rahilly, R. & Tucker, J. A. (1973). The early development of the larynx. Embryos of the first five weeks. *Annals of Otology, Rhinology and Laryngology*, Supplement 71.

Reidenberg, J. S. & Laitman, J. T. (1991). Effect of basicranial flexion on larynx and hyoid in rats. *Anatomical Record*, **230**, 557–69.

Shikinami, J. (1926). Detailed form of the Wolfian body in human embryos of the first eight weeks. *Contributions to Embryology, Carnegie Institute*, **8**, 46–61.

Soulie, A. & Bardier, E. (1907). Recherches sur le developpment du larynx chez l'homme. *Journal Anatomy et Physiology*, **43**, 137–240.

Thorogood, P. & Ferreti, P. (1992). Heads and Tales: recent advances in craniofacial development. *British Dental Journal*, **173**, 301–6.

Walander, A. (1950). Prenatal development of the epithelial premordium of the larynx in rats. *Acta Anatomica*, **10**, Supplement 13.

Westhorpe, R. N. (1987). The position of the larynx in children and its relationship to the ease of intubation. *Anaesthetic Intensive Care*, **15**, 384–8.

Wind, J. (1970). *On the Phylogeny and the Ontogeny of the Human Larynx*. Groningen, Walters-Noordhoff.

Wolfson, V. P. & Laitman, J. T. (1990). Ultrasound investigation of foetal upper respiratory tract anatomy. *Anatomical Record*, **227**, 363–72.

Zaw-Tun, H. A. & Burdi, A. R. (1985). Reexamination of the origin and early development of the human larynx. *Acta Anatomica*, **122**, 163–84.

4.2 Framework, cartillages, ligaments and joints

Ardran, G. M. & Kemp, F. H. (1966). The mechanism of the larynx. *British Journal of Radiology*, **39**, 649–53.

Audemorte, T. B., Sheridan, P. S. & Holt, R. G. (1983). Autoradiographic evidence of sex steroid receptors in laryngeal tissues of the baboon (*Papio cynocephalus*). *Laryngoscope*, **93**, 1607–11.

Beckford, N. S., Rood. S, R. & Schaid, D. (1985). Androgen stimulation and laryngeal development. *Annals of Otology, Rhinology and Laryngology*, **94**, 634–40.

Cohen, S. R., Perelman, N. & Nimni, M. E. (1993). Whole organ evaluation of collagen in the developing human larynx and adjoining anatomic structures (hyoid and trachea). *Annals of Otology, Rhinology and Laryngology*, **1102**, 655–9.

Denny, S. P. (1976). Comparative anatomy of the larynx. In *Scientific Basis of Otolaryngology*, ed. R. Hinchcliffe & D. F. N. Harrison, p. 536. London, Heinemann.

Fink, R. B. & Demarest, R. J. (1978). *Laryngeal Biomechanics*. Harvard University Press.

Harrison, D. F. N. (1983). Correlation between clinical and histological staging in laryngeal cancer. PhD thesis, University of London.

Harrison, D. F. N. & Denny, S. P. (1983). Ossification in the primate larynx. *Archives of Otolaryngology*, 95, 440–6.

Harrison, D. F. N. & Denny, S. P. (1985). The possible influence of laryngeal and tracheal size on the running speed of mammals. *Acta Otolaryngologica (Stock.)*, 99, 229–35.

Hately, W., Evison, G. & Sameul, E. (1965). The pattern of ossification in laryngeal cartilage: a radiological study. *British Journal of Radiology*, 38, 585–90.

Hirano, M., Kurita, S. & Yukizane, K. (1989). Asymmetry of the laryngeal framework: a morphological study of cadaver larynges. *Annals of Otology, Rhinology and Laryngology*, 98, 135–40.

Holt, R. G., Aufemorte, T. B. & Sheridan, P. S. (1986). Estrogen receptors in the larynx of the aged baboon (*Papio cynocephalus*). *Annals of Otology, Rhinology and Laryngology*, 95, 608–17.

Ibrahim, I. A. & Youssif, M. J. (1991). Comparative anatomical and metrical studies on the laryngeal cartilages of goat and sheep. *Asian Veterinary Medical Journal*, 50, 1–16.

Keen, J. A. & Wainwright, J. (1958). Ossification of the thyroid, cricoid and arytenoid cartilage. *South African Journal of Laboratory and Clinical Medicine*, 4, 83–90.

Maue, W. M. & Dickson, D. R. (1971). Cartilages and ligaments of the adult human larynx. *Archives of Otolaryngology*, 94, 432–9.

Milroy, C. M. (1992). Ossification of the Epiglottis. *Journal of Laryngology and Otology*, 106, 180–2.

Pracy, R. (1984). The functional anatomy of the immature mammalian larynx. A study with comparative anatomy of the herbivore, carnivore and primate larynx. MPhil thesis, University of London.

Reiss, M. (1982). Males bigger, females biggest. *New Scientist*, October 28, 226–9.

Saber, A. S. (1983). The cartilages of the larynx of the one-humped camel (*Camelus dromedarius*) and sheep (*Ovis aries*). *Anatomica Histologica and Embryologica*, 12 (1), 77–84.

Schild, J. A. (1984). Relationships of laryngeal dimensions to body size and gestational age in premature neonates and small infants. *Laryngocope*, 94, 1284–92.

Sellars, I. & Keen, E. N. (1978). The anatomy and movements of the cricoarytenoid joint. *Laryngoscope*, 88, 667–74.

Sellars, I. & Sellars, S. (1983). Cricoarytenoid joint structure and function. *Journal of Laryngology and Otology*, 97, 1027–34.

Solis, J. A. (1976). Comparative features of the laryngeal skeleton of the Phillipine water buffalo. *Phillipine Journal of Veterinary Medicine*, 15, 21–7.

Tuohimaa, P. T., Kallio, S. & Heinijoki, J. (1981). Androgen receptors in laryngeal cancer. *Acta Otolaryngologica (Stock.)*, 91, 159–54.

Vastine, J. H. & Vastine, M. R. (1952). Calicification in the laryngeal cartilage. *Archives of Otolaryngology*, 55, 1–8.

Williams, R. G. & Eccles, R. (1990). A new clinical measure of external laryngeal size which predicts fundamental frequency of the larynx. *Acta Otolaryngologica (Stock.)*, 110, 141–8.

Wind, J. (1976). Phylogeny of the human vocal tract. *Annals of the New York Academy of Medicine*, 2280, 612–30.

Zrunek, M., Happak, W., Herrman, M. & Streinzer, W. (1988). Comparative anatomy of human and sheep laryngeal skeleton. *Acta Otolaryngologica (Stock.)*, 105, 155–62.

4.3 Ventricle, saccule and air sac

Birt, D. (1987). Observations on the size of the saccule in laryngectomy specimens. *Laryngoscope*, 97, 190–200.

Broyles, E. N. (1959). Anatomical observations concerning the laryngeal appendix. *Annals of Otology Rhinology and Laryngology*, 68, 461–70.

Canalis, R. F. (1980). Laryngeal ventricle, historical features. *Annals of Otology, Rhinology and Laryngology*, 89, 184–7.

Delahunty, J. E. & Cherry, J. (1969). The Laryngeal Saccule. *Journal of Laryngology and Otology*, 83, 803–15.

Denny, S. P. (1976). Comparative anatomy of the larynx. In *Scientific Basis of Otolaryngology*, ed. R. Hinchliffe & D. F. N. Harrison, p. 536. London, Heinemann.

De Santo, L. W. (1974). Laryngocoele, laryngeal mucocoele, large saccules and larynygeal saccular cyst: a developmental spectrum. *Laryngoscope*, 84, 1291–4.

Fabricius, H. (1600). *De Vione, Voce, Aditu*. Venetiis.

Garrison, D. H. & Hast, M. H. (1993). Andreas Vesalius on the larynx and hyoid bone. An annotated translation from the 1543 and 1555 editions of *De Humani Corporis Fabrica*. *Medical History*, 37, 3–36.

Harrison, D. F. N. (1983). Correlation between clinical and histological staging in laryngeal cancer. PhD thesis, University of London.

Hast, M. (1970). Developmental anatomy of the larynx. *Otolaryngological Clinics of North America*, 3, 413–38.

Hilton, J. (1837). Description of the sacculus or pouch in the human larynx. *Guy's Hospital Reports*, 2, 510–25.

Kotby, M. N., Kirchner, J. A. & Basiouny, S. E. (1991). Histo-anatomical structure of the human laryngeal ventricle. *Acta Otolaryngologica (Stock.)*, 111, 396–402.

Larrey, D. J. (1829). Clinique surgicale exercee particulierement dans les camps et les hospitaux militaires depuis. *Clinical Chirugurie*, 2, 81–2.

Martin, R. D. (1990). *Primate Origins and Evolution*, p. 716. London, Chapman & Hall.

Morgan, E. (1990). *The Scars of Evolution*. London, Souvenir Press.

Napier, J. R. & Napier, P. H. (1985). *The Natural History of Primates*. London, The British Museum (Natural History).

Negus, V. E. (1949). *The Comparative Anatomy and Physiology of the Larynx*. London, Heinemann.

Santorini, G. D. (1724). *Observations Anatomical*, pp. 106–249. Venetiis.

Schmidt-Neilsen, K. (1984). *Scaling: Why is Animal Size So Important?* p. 100. Cambridge, Cambridge University Press.

Scott, G. B. D. (1976). A morphometric study of the laryngeal saccule and sinus. *Clinical Otolaryngology*, 1, 115–22.

Stark, D. & Schneider, R. (1960). Respiratinsorgane larynx. In *Primatologica*, ed. H. Hofer, A. H. Schultz, & D. Stark, pp. 124–40. Basel, ˜Larger.

Stell, P. M. & Maran, A. G. D. (1975). Laryngocoele. *Journal of Laryngology and Otology*, 89, 915–24.

Virchow, R. (1867). *Die Krankhaften Geschwultste*, 3–35. Berlin.

Young, B. A. (1992). Tracheal diverticula in snakes, possible functions and evolution. *Journal of Zoology (London)*, 227, 567–83.

Zemlin, W., Elving, S. & Hull, L. (1984). The superior thyrohyoid muscle in the human larynx. *American Speech and Hearing Association*, 26, 71–4.

4.4 Laryngeal mucous glands

Altmann, F., Ginsberg, I. & Stout, A. P. (1952). Intraepithelial carcinoma (carcinoma *in situ*) of the larynx. *Archives of Otolaryngology*, 56, 121–33.
Bridger, G. & Nassar, V. H. (1971). Carcinoma *in situ* involving the laryngeal, mucous glands *Archives of Otolaryngology*, 94, 389–400.
Fink, B. R. & Beckwith, J. B. (1980). Laryngeal mucous gland excess in victims of Sudden Infant Death. *American Journal of Diseases of Children*, 134, 144–6.
Gregson, R. (1989). All night–every night. *British Medical Journal*, 299, 1049–50.
Harrison, D. F. N. (1983). Correlation between clinical and histological staging in laryngeal cancer. PhD thesis, University of London.
Harrison, D. F. N. (1991). Laryngeal morphology in sudden unexpected death in infants. *Journal of Laryngology and Otology*, 1105, 646–50
Nassar, V. H. & Bridger, P. (1971). Topography of the laryngeal mucous glands. *Archives of Otolaryngology*, 94, 490–8
Pracy, R. (1984). The functional anatomy of the immature mammalian larynx. A study with comparative anatomy of the herbivore, carnivore and primate larynx. MPhil thesis, University of London.
Scott, G. B. D. (1976). A quantitative study of microscopical changes in the epithelium and subepithelial tissue of the laryngeal folds, sinus and saccule. *Clinical Otolaryngology*, 1 257–64
Stell, P. M., Gudrun, R. & Watt, J. (1981). Morphology of the human larynx. III. The Supraglottis. *Clinical Otolaryngology*, 6, 389–93

4.5 Musculature

Arnold, G. E. (1961). Physiology and pathology of the cricothyroid muscle *Annals of Otology, Rhinology and Laryngology*, 71, 687–753.
Fink, B. R. (1973). The curse of Adam: effort closure of the human larynx. *Anesthesiology*, 3, 325–7.
Fink, B. R. (1978). Energy and the larynx. *Annals of Otology, Rhinology and Laryngology*, 87, 1–11.
Harrison, D. F. N. & Denny, S. P. (1985). The possible influence of laryngeal and tracheal size on running speed in mammals. *Acta Otolaryngologica*, 99, 229–35.
Hast, M. H. (1968). Studies of the extrinsic laryngeal muscles. *Archives of Otolaryngology*, 88, 273–8.
Hast, M. H. (1969). The primate larynx. *Acta Otolaryngologica (Stock.)*, 67, 84–92.
Henle, J. (1873). *Hands Eingeweldelehre*.
Hetherington, J. (1934). The kerato-cricoid muscle in the American white and negro. *American Journal of Physical Anthropology*, 19, 203–12.
Konrad, H. R., Rattenberg, C. C., Kahn, M. L. *et al.* (1984). Opening and closing mechanisms of the larynx. *Otolaryngology – Head and Neck Surgery*, 92 (4) 402–5.
Kotby, M. N. & Haugen, L. K. (1970a). The mechanics of laryngeal function *Acta Otolaryngologica (Stock.)*, 70, 203–11.
Kotby, M. N. & Haugen, L. K. (1970b). Attempts at evaluation of the function of various laryngeal muscles in the light of muscle and nerve stimulation experiments in man. *Acta Otolaryngologica (Stock.)*, 70, 419–27.

Kotby, M. N., Kirchner, J. A., Kahane, J. C. *et al.* (1991). Histo-anatomical structure of the human laryngeal ventricle. *Acta Otolaryngologica (Stock.)*, 111, 396–402.

Malmgren, L. T. & Gacek, R. R. (1981). Histochemical characterisics of muscle fiber types in the posterior cricoarytenoid muscle. *Annals of Otology, Rhinology and Laryngology*, 90, 423–9.

Mayet, A. & Muendnich, K. (1958). Beitrag zur anatomie und zur Funkton des M. Cricthyroideus und der cricothroidgelenke. *Anatomy anz Anatomica*, 33, 273–85.

Morgagni, J. B. (1741) 'Adversaria Anatomia'. Anat Lugd Batar, L, XV1.

Mossallam, I., Kotby, M. N., Abd-El-Rahman. & El-Samma. (1987). Attachment of some internal laryngeal muscles at the base of the arytenoid cartilage. *Acta Otolaryngologica (Stock.)*, 103, 649–56.

Sahgal, V. J. & Hast, M. H. (1974). Histochemistry of primate laryngeal muscles. *Acta Otolaryngologica (Stock.)*, 78, 277–81.

Santorini, G. D. (1724). Observations Anatomical, pp. 106–249. Venetiis.

Sellars, I. E. (1978). A re-appraisal of intrinsic laryngeal muscle action. *Journal of Otolaryngology*, 7, 450–6.

Sharp, J. F. (1990). The ceratocricoid muscle. *Clinical Otolaryngology*, 15, 257–61.

Turner, W. M. S. (1860). Remarks on musculus kerato-cricoideus. *Edinburgh Medical Journal*, 5, 744–6.

Vesalius, A. (1545). *De Corporis Humani Fabrica*.

Zaretsky, L. & Sanders, I. (1992). The three bellies of the canine cricothyroid muscle. *Annals of Otology, Rhinology and Laryngology*, (Suppl 156), 101 1–16.

Zemlin, W., Elving, S. & Hull, L. (1988). The superior thyroarytenoid muscle in the human larynx. *American Speech and Hearing Association*, 26, 71–8.

Zrunek, H. M., Pechmann, U. J. & Streinzer, W. (1989). Comparative histochemistry of human and sheep laryngeal muscles. *Acta Otolaryngologica (Stock.)*, 107, 283–8.

4.6 Nerve supply

Abo-El-Enein (1967). Functional anatomy of the larynx. PhD thesis, University of London.

Adzaku, F. K. & Wyke, B. (1979). Innervation of the subglottic mucosa of the larynx and its significance. *Folia Phoniatricia*, 31, 271–83.

Barson, A. J. (1970). The vertebral level of termination of the spinal cord during normal and abnormal development. *Journal of Anatomy*, 106, 489–97.

Berry, J. (1888). Suspensory ligaments of the thyroid gland. *Journal of Anatomy and Physiology*, 22, iv–v.

Bowden, R. E. M. & Schuer, J. L. (1961). Innervation of the larynx. Comparative studies of the nerve supply in Eutherian mammals. *Proceedings of the Zoological Society of London*, 136, 325–30.

Cook, W. R. (1976). Idiopathic Laryngeal Paralysis in the Horse, a clinical and pathological study with particular reference to Diagnosis, Aetiology and treatment. PhD thesis, University of Cambridge.

Cook, W. R. & Kirk, W. (1991). Hereditary diseases of the horse and their prevention. *Irish Veterinary Journal*, 44, 59–66.

Crumley, R. L. (1982). Experiences in laryngeal nerve reinnervation. *Laryngoscope* (Supplement 30), 92 (9).

Cunningham, P. (1991). The genetics of the thoroughbred horse. *Scientific American*, May, 56–62.

Dahlqvist, A., Carlsoo, B. & Hellstrom, S. (1982). Fiber components of the recurrent laryngeal nerve of the rat. *Anatomical Record*, **204**, 365–70.

Dahlqvist, A. & Forsgren, S. (1989). Networks of peptide-containing nerve fibres in laryngeal nerve paraganglia. *Acta Otolaryngologica (Stock.)*, **107**, 289–95.

Derkson, F. J., Stick, J. A., Scott. E. A. *et al.* (1986). Effect of laryngeal hemiplegia and laryngoplasty on airway flow mechanics in exercising horses. *American Journal of Veterinary Research*, **47**, 16–19.

Dilworth, T. F. M. (1921). The nerves of the human larynx. *Journal of Anatomy*, **56**, 48–52.

Duncan, I. D., Griffiths, I. R., McQueen, A. & Baker, G. O. (1974). The pathology of equine laryngeal hemiplegia. *Acta Neuropathologica*, **27**, 337–42.

Evans, D. H. L. & Murray, J. G. (1954). Histological and functional studies on the fibre composition of the vagus nerve in the rabbit. *Journal of Anatomy*, **88**, 320–37.

Fraher, J. P. (1980). On methods of measuring nerve fibres. *Journal of Anatomy*, **130**, 139–51.

Gacek, R. R. & Lyon, M. J. (1976). Fibre components of the recurrent laryngeal nerve in the cat. *Annals of Otology, Rhinology and Laryngology*, **85**, 460–71.

Harrison, D. F. N. (1981). Fibre size frequency in the recurrent laryngeal nerves of man and giraffe. *Acta Otolaryngologica (Stock.)*, **91**, 383–9.

Hauser-Kronberger, C., Albegger, K. & Saria, A. (1993). Regulating peptides in human larynx. *Acta Otolaryngologica (Stock.)*, **113**, 4409–13.

Hisa, Y., Uno, T., Tadaki, N., *et al.* (1992). Distribution of calcitonin gene-related peptide nerve fibres in the canine larynx. *European Archives of Otolaryngology*, **249**, 52–5.

Kleinsasser, O. (1964). Das Glomus Laryngicum Inferior. *Archives Ohren-Nasen und Kehlkopfhedk*, **184**, 214–24.

Lang, J., Nachbaur. T., Fuscher, K. & Vogel, B. (1987). Uber den Nervus laryngeal superior und die aterial laryngeal superior. *Acta Anatomica*, **130**, 309–18.

Lemere, F. (1932). Innervation of the larynx. *American Journal of Anatomy*, **51**, 417–37.

Lopez-Plana, C., Sautet, J. T. & Pons, J. (1989). Morphometric study of the recurrent laryngeal nerve in 'normal horses'. From the Laboratory of Anatomy, Ecole Nationale Veterinaire, Toulouse, France.

Martensson, H. & Terins, J. (1985). Recurrent laryngeal palsy in thyroid gland surgery related to operations and nerves at risk. *Archives of Surgery*, **120**, 475–77.

Mayhew, T. M. (1983). Stereology progress in qualitative microscopical anatomy. In *Progress in Anatomy*, vol. 3, pp. 81–112. Cambridge, Cambridge University Press.

Mayhew, T. M. & Sharma, A. K. (1984). Sample schemes for estimating nerve fibre size. *Journal of Anatomy*, **139**, 45–58.

Murray, J. G. (1957) Innervation of the intrinsic muscles of the cat's larynx by the recurrent laryngeal nerve: a unimodal nerve. *Journal of Physiology*, **135**, 206–12.

Murtagh, J. A. & Campbell, C. J. (1951). The relation of fibre size to function in the recurrent laryngeal nerve. *Laryngoscope*, **61**, 581–90.

Nguyen, M., Junien-Lavillauroy, C., Faure, C. (1989). Anatomical intra-laryngeal anterior branch study of the recurrent laryngeal nerve. *Surgical Radiological Anatomy*, **11**, 123–7.

O'Reilly, P. M. & Fitzgerald, M. J. T. (1985). Internodal segments in human laryngeal nerves. *Journal of Anatomy*, **140**, 645–50.

Peytz, F., Rasmussen, H. & Buchthal, F. (1965). Conduction time and velocity in human recurrent laryngeal nerves. *Danish Medical Bulletin*, **12**, 125–7.

Ramaswamy, S. & Kulasekaran, D. (1974). The ganglion on the internal laryngeal nerve. *Archives of Otolaryngology*, 100, 28–31.

Reeve, T. S., Coupland, G. A. E. & Johnson, D. L. (1969). The recurrent and external nerves in thyroidectomy. *Medical Journal of Australia*, 1, 380–2.

Salama, A. B. & McGrath, P. (1992). Recurrent laryngeal nerve and the posterior fascial attachment of the thyroid gland. *Australian and New Zealand Journal of Surgery*, 62, 444–9.

Saunders, J. B. & Malley, C. D. O. (1982). *The Anatomical Drawings of Andrew Versalius*, p. 152. New York, W. B. Saunders.

Scheur, J. L. (1964). Fibre size frequency distribution in normal human laryngeal nerves. *Journal of Anatomy*, 98, 99–104.

Shin, T. & Rabuzzi, D. (1971). The vertebral level of termination of the spinal cord during normal and abnormal development. *Journal of Anatomy*, 106, 489–97.

Steinberg, J. L., Khane, G. J. & Fernandes, C. M. C. (1986). Anatomy of the recurrent laryngeal nerve: a rediscription. *Journal of Laryngology and Otology*, 100, 919–27.

Tomasch, J. & Britton, W. A. (1955). A fibre analysis of the laryngeal nerve supply in man. *Acta Anatomica*, 23, 386–98.

Vogel, P. H. (1952). The innervation of the larynx in man and dog. *American Journal of Anatomy*, 90, 427–47.

Williams, A. F. (1951). The nerve supply of the laryngeal muscles. *Journal of Laryngology and Otology*, 64, 243–8.

Wyke, B. D. & Kirchner, J. A. (1976). Neurology of the Larynx. In *Scientific Basis of Otolaryngology*, ed. R. Hinchcliffe and D. F. N. Harrison, Chapter 40. London, Heinemann.

Yoshikazu, Y., Shimazaki, T. & Tanaka, Y. (1993). Ganglions and ganglionic neurones in the cat larynx. *Acta Otolaryngologica*, 113, 415–20.

4.7 Vascular supply

Andrea, M. (1975). Vascularizacaco arterial da laringe. Candidates dissertation for Doctorate in Faculty of Medicine, University of Lisbon.

Andrea, M. (1981). Vasculature of the Anterior Commissure. *Annals of Otology, Rhinology and Laryngology*, 90, 18–20.

Harrison, D. F. N. (1971). Surgical anatomy of the maxillary and ethmoid sinuses. *Laryngoscope*, 81, 1658–64.

Hirano, M. & Sato, K. (1993). *Histological Color Atlas of the Human larynx*, p. 106–7. San Diego, Singular Publishing Group.

Pearson, B. W. (1975). Laryngeal microvasculature and pathways of cancer spread. *Laryngoscope*, 85, 700–13.

Terracol, J. & Guerrier, Y. (1951). Le systeme arterial du larynx. *Etude Anatomique Montpelier Medicale*, 41/42, 340–65.

5 Laryngeal physiology

5.1 Respiration

Whilst the study of structure is best discussed under the general heading of morphology, the function of the living organism is considered under the heading of physiology. This is about food, digestion, circulation, respiration and all functions which maintain an animal's fitness for survival. To understand how this is effected it is necessary to be familiar not only with the structure of individual species but also the means by which they adjust to environmental pressures. A comparative approach, examining how mammals solve the hazards of living within the constraints of variable environments, provides some insight into our understanding of the correlation and integration of function necessary for existence. No mammal is independent of its environment and all must be able to cope with its variabilities and challenges.

Comparative mammalian physiology is important as an intellectual discipline in its own right, but it provides the perspective that is vital to the understanding of physiological processes as they occur in individual species. This chapter is primarily concerned with the functional role of the larynx in protection of the lower respiratory tract, respiration, locomotion and vocalization. This unique organ can best be appreciated by recognition of its origin which was determined by primitive needs. In an aquatic environment the primitive larynx functioned as a simple sphincter designed to protect the lower airway from intrusion of water and other foreign matter. Dilation of this sphincter became necessary for existence in a terrestrial environment, some amphibians phylogentically acquiring lateral laryngeal cartilages into which were inserted dilator muscles. Eventually, cartilaginous rings between glottis and trachea in some higher vertebrates (such as alligators) developed allowing controlled aerial respiration. Comparative observations of modern amphibians suggest that aerial respiration in early land-dwelling vertebrates was accessory to gill and skin respiration. Functioning gills are present in the early stages of all amphibians, and in anuran amphibians such as the frog, the larynx consists of a pair of arytenoid cartilages together with a circular cricoid cartilage. Reptiles show a similar laryngeal structure to anuran amphibians with both dilator and constrictor muscles. These modifications are for controlling the passage of air (and contained oxygen) to the lungs, which exemplify the importance of the larynx in mammalian respiration.

Much of the data relating to neuromuscular activity within the larynx has been obtained from animal experimentation because of the constraints inherent in applying invasive techniques to humans. The basic functions of the larynx (protective, respiratory and phonatory) are controlled by a complex interrelationship of diverse polysynaptic brainstem reflexes clearly summarized by Sasaki & Buckwalter (1981). Whilst protective function in all mammals is entirely reflex and therefore involuntary, respiratory and phonatory activities may be initiated voluntarily through an array of 'feedback' reflexes. The protective function of the larynx may be examined neurophysiologically by studying the glottic closure reflex. This reflex produces closure during deglutition; in humans the threshold of the adductor reflex measures 0.5 V and has a latency of 25 msec. This finding shows that this is a polysynaptic brainstem reflex. However, unlike many other mammals humans do not have a crossed adductor reflex, that is, stimulation of one superior laryngeal nerve does not produce simultaneous action potentials in the contralateral adductor musculature. It is possible that such interspecial differences are secondary to variations in distribution of the laryngeal nerves, previously described in section 4.6. Sphincteric closure in humans produced by bilateral stimulation of the superior laryngeal nerves results in adduction of three muscular layers within the laryngeal framework. The highest, or first, level of closure occurs at the aryepiglottic folds, which contain the superior division of the thyroarytenoid muscle. Reflex contraction allows the folds to approximate, covering the superior aspect of the laryngeal inlet. The anterior gap is filled by the epiglottis and posteriorly, the arytenoid cartilages complete closure. Although this tier is of considerable importance in humans where the larynx lies low in the neck, it may be of less importance in mammals with an elongated intranarial epiglottis with or without high aryepiglottic folds. This anatomical arrangement allows for almost simultaneous feeding, breathing and olfactory sampling. A similar configuration is found in the human until the age of 2 years by which time the larynx has descended to its adult position. During deglutition the larynx is elevated by the pull of muscles attached to or originating from the hyoid bone. This results in deflection of food or liquid into lateral channels within the pyriform fossae before entering the oesophagus. This elevation also results in displacement of the epiglottis inferiorly towards the closed aryepiglottic folds, enhancing protection of the laryngeal lumen.

The second level of protection is applicable to mammals possessing a laryngeal ventricle and false cords. Contraction of the thyroarytenoid muscle allows bilateral approximation of the false cords, which in most species are directed inferiorly at their free edges preventing air escape during coughing. The pressure required to break this 'passive' valvular effect was estimated as 30 mm Hg in cadaver larynges (Bruton, 1885).

The third layer in all mammals is at the level of the vocal cords which show interspecies variation in their orientation and thickness. The inferior division of the thyroarytenoid muscle forms the main bulk of this 'shelf', which in humans is upturned with the capability for tight closure. The pressure required to overcome this valvular effect approaches 140 mm Hg. The thyroarytenoid muscle is one of the most rapidly contracting striated muscles (approaching the medial rectus muscle of the eye in speed of contraction) and in dogs contracts in 14 msec. This compares with 40 msec for the cricothyroid and 44 msec for the posterior cricothyroid muscles (Hast, 1967).

The diverse functions of the larynx range from protective glottic closure to the fine motor control needed for phonation. These are co-ordinated by a complex system of afferent and efferent nerve fibres related to brainstem mediated reflex arcs, which have significant cortical connections. The afferent system is derived from branches of both divisions of the superior laryngeal nerves, recurrent laryngeal nerves and, when present, communicating branches. Fukuyama et al. (1993) detected evoked potentials on the dural surface of cat brains by electrical stimulation of the superior laryngeal nerve. They called this the laryngeal sensory evoked potential (LSEP) and found it to consist of five components: N1, N2, N4, N12 and a large biphasic potential which lasted 6 to 18 msec. Although the glottic closure reflex can be elicited by direct superior laryngeal nerve stimulation, in the normal state this reflex is triggered via stimulation of tactile receptors present on the laryngeal surface of the epiglottis and in the arytenoid region. Mucosal stimulation of these regions in the cat by Fukuyama et al. (1993) elicited similar morphological, and neurophysiological responses to those found on superior laryngeal nerve stimulation.

Excitation of most major cranial afferent nerves produces a strong adductor response and, in the cat, reflex potentials in this muscle can be elicited by stimulation of optic, acoustic, trigeminal, vagus and even intercostal nerves. Receptors in the subglottic mucosa have also been shown to produce glottic closure in most mammals and the susceptibility of this response to such diverse sensory stimulation serves to emphasize its primitive role in protection of the lower respiratory tract.

Role of the larynx in respiration

A fundamental requirement for survival in any mammal is energy metabolism, which itself is closely linked to respiration. For terrestrial animals the atmosphere contains adequate supplies of oxygen even at high altitudes, whilst mammals that have adapted to living in water, such as seals and whales, have oxygen only available at the water surface. Aerobic respiration is the fundamental process by which mammals, of diverse form and behaviour, take in oxygen and dispose of carbon dioxide. The means by which they satisfy their requirements is complex, although all subserve a relatively unified pattern of cellular metabolism. It is customary to divide discussion of respiration into two distinct parts: internal and external. Internal respiration is defined as the sum of all enzymatic reactions within the cell by which energy is made available, and external respiration is concerned with the mechanisms by which the animal obtains oxygen and removes carbon dioxide. It is the vertebrates who have developed air-breathing lungs with mechanisms for regular ventilation. In all mammals the vocal cords act as a regulatory obstruction to the flow of air. The control of airway resistance at this level is of paramount importance in the understanding of comparative behaviour.

General considerations of comparative respiration

The amount of oxygen in the earth's atmosphere is remarkably constant at 20.95% with the pO_2 in dry air at sea level with a barometric pressure of 760 mm being 159.2 mm. With increasing altitude barometric pressure falls and at 5000 m the pO_2 will have fallen to 88 mm, with a much diminished available pO_2 at the pulmonary side of the respiratory epithelium (Jones, 1972). Maintenance of the highest possible oxygen pressure at the respiratory epithelial interface is desirable but may not always be possible because of morphological reasons such as a large dead space within the trachea. Of greater physiological importance is oxygen utilization, defined as the 'percentage withdrawal of oxygen from the respiratory cycle' and designated by the difference in oxygen content between pulmonary arterial and venous blood. Some indication of this can be reached by comparing oxygen content between inspired and expired air, giving a value for humans at rest of approximately 22%. Oxygen need is related, primarily, to an animal's energy requirements with greater demand for those that are large and active. Although this can be met to a variable degree by anaerobic respiration (as in the cheetah's 200 m sprint) there is great advantage to be gained by maximal use of aerobic processes, for which there are many variations and adaptations.

Measurement of oxygen consumption in many mammals is relatively easy,

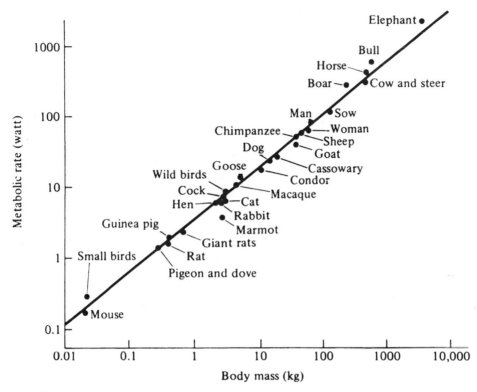

Figure 5.1. Metabolic rates for mammals and birds, when plotted against body mass on logarithmic coordinates. (After Schmidt Nielson, 1984.)

but it may not be commensurate with the metabolic rate. Anaerobic respiration, depending on non-oxidative metabolism, has zero oxygen consumption and a definitive metabolic rate. For the more common aerobic respiration the rate of oxygen consumption (measured as oxygen uptake per unit time) represents the animal's metabolic rate. In the resting state, the rate of oxygen consumption (viz. the metabolic rate) is recognized as related to body size (Figure 5.1). This is not a direct relationship, for as Rubner (1883) showed in one breed of dog, oxygen uptake is proportional to two-thirds of the power of the body weight, which is a close estimate of surface area. This relationship was quantified by Kleiber (1932) following measurement of the metabolic rate in 4000 animals, ranging in size from rats to steers. By expressing metabolic rate in kcal day^{-1} and body weight in kg he produced an allometric equation:

$$P_m = 73.3\, M_b^{0.74}$$

This corresponded to a data plot on logarithmic coordinates that gave a straight regression line with a slope of 0.74. If the metabolic rate were a direct function of body weight (or surface area) the slope would have been 0.67.

In his discussion of the structural and functional consequences of changes in size or scale amongst similar organisms, Schmidt-Nielsen (1984) makes use of allometric equations to scale morphological and physiological variables. These equations are based on the knowledge that when two variables are plotted on logarithmic coordinates the result is a straight line. The proportional coefficient a (the intercept at unity) and the exponent b (the slope of the regression line) have different meanings and can answer different questions. For example, the equations for the metabolic rates of marsupials and eutherian mammals have similar exponents but the proportional coefficients are lower for marsupials. It may therefore be considered that in general, marsupials have lower metabolic rates than eutherian mammals. The exponent, however, suggests that metabolic rate varies with changing body size similarly in both marsupials and eutherian mammals. Statistics are no more than descriptive numbers and significant statistics do not necessarily equate with biological significance!

It has long been assumed that the slow rate of growth of young marsupials was related to their 30% lower metabolic rates than that found in eutherians. Thompson and Nicoll (Loudon, 1986) however, found that during reproduction marsupial metabolic rates equalled those found in eutherians. This was surprising since the latter usually show no change during breeding. They suggested that a high resting metabolic rate could be advantageous or even necessary for mammalian reproduction and that species with normally low rates would show an increase when breeding. Other investigators found that although many large grazing eutherians had high metabolic rates, others were low outside of the breeding season, being similar to marsupials (Loudon, 1986). Since mammals show seasonal patterns of feeding, growth and reproduction their environment can influence major changes in their resting metabolic rate, particularly in response to day length. In spite of these reservations it is generally accepted that the body weight exponent is close to 0.75, although lack of data prevents sufficient understanding of the principles that underlie this slope, particularly when considering the wide range of mammals that exist in varying environments. For example, shrews tend to have very high metabolic rates (and short lives) and their data deviate from the general mammalian regression line. The Etruscan shrew (*Suncus etruscus*) with a body weight averaging 2 g needs to eat hourly, consuming six times its body weight in insects each day. Its oxygen consumption averages 0.4 ml O_2 min g^{-1} body mass, which is six times the maximal weight specific oxygen consumption in humans (Morrison *et al.*, 1959).

Maximal oxygen consumption is proportional to the mass of mitochondria in skeletal muscle and is capable of quantitative adaption with changes in energy demand (Weibel, 1979). Consequently, the high mass-specific metabolic needs of this small mammal may be related to the size of the mitochondrial content of its muscle cells, as well as a relative increase in the size of the pulmonary gas exchange mechanism.

The mitochondria in the muscles of a trained human runner can utilize more than 5 litres of oxygen each minute. To maintain metabolism at this rate a continuous flow of oxygen must pass from environmental air to the lungs by inspiration. Following transference to the pulmonary circulation and carriage by the blood erythrocytes, it finally reaches the mitochondria by diffusion. Conversely, carbon dioxide produced during metabolism has to be removed along similar pathways finally being removed from the lungs on expiration.

Blood and gas transportation

For most mammals the supply of oxygen is more important than elimination of carbon dioxide. The principle parameters in oxygen carriage are haemoglobin concentration, blood volume and size of red cells. Burke (1966) measured the haemoglobin concentration in mammals of varying size finding a mean value of 128.7 g per litre of blood with no relation to body size. The associated oxygen carrying capacity was 175 ml per litre, but this did not include aquatic mammals. More recent data suggest that the haemoglobin concentration for most mammals does not exceed 150 g per litre of blood and is independent of body size. Small mammals such as shrews with their high metabolic rates and demand for oxygen appear to have blood with similar oxygen carrying capacity as large mammals. Although haemoglobin concentration in mammals appears to be scale-independent, special physiological demands lead to levels in excess of these average figures. Animals such as the yak (*Bas mutus*) and lama (*Lama lama*) who live at high altitudes have much higher haemoglobin levels, whilst in some insectivorous bats this may reach 244 g per litre of blood (Jurgens et al., 1981). Such an increase in haemoglobin concentration leads to an increase in blood viscosity and cardiac effort. Blood volume in most mammals is between 60 and 70 cm^3 blood per kilogram body mass. Stahl (1965) calculated the equation for blood volume relative to body mass as being:

$$V_b = 65.6 \ M_b^{1.02}$$

Since the exponent is close to 1.0, it appears that the blood volume is a constant fraction of body mass. However, exceptions are to found in diving mammals who have much larger than calculated blood volumes, oxygen being stored in the plasma (this is discussed later in section 5.2). Blood volume can be estimated by a number of sophisticated techniques, although most data have been obtained by bleeding with measurement of residual blood after maceration and extraction. This presents formidable problems in larger mammals but appears to range between 6 and 10% of body weight.

Similar agreement is found in red blood cell size which ranges for most mammals between 5 and 8 μm. All are round, biconcave discs and Altman & Dittmer (1961) measuring diameters in more than 100 mammals found the largest to be in the elephant (9.2 μm) and the and the smallest in sheep (4.8 μm). Schmidt-Nielsen (1984) concluded from these figures that the capillaries of all mammals must be of a similar diameter, and that the high metabolic rates in some small mammals do not require special adaptions in red cell or carrying capacity.

Comparative morphology of the trachea and lung

Those parts of the respiratory tract that are ventilated but play no role in gas exchange with the pulmonary capillary blood make up the respiratory 'dead space'. In humans this is about 140 ml, so that 28% of the 500 ml of air inspired in the resting state fails to reach the area of gaseous exchange. Increased oxygen demand can increase the volume of inspired air to over 4000 ml but at the extreme limit of expiration there remains a residual volume of more than 1000 ml, exclusive of dead space. This 'used', air will be mixed with fresh inspired air thus lowering the alveolar composition to less than 150 mm pO_2 (Jones, 1972). The importance of the dead space in mammalian respiration has been evaluated by Tenny & Bartlett (1967) who examined animals ranging from shrews to whales, finding that the tracheal dead space was close to a constant fraction of lung volume. They concluded that the dead space would therefore have a similar relative effect on lung ventilation in all mammals. This may not be so in ruminants such as the giraffe. Rumen gas contains not only nitrogen but varying amounts of methane, oxygen and carbon dioxide. Rate of gas production in the giraffe is variable but can reach 0.7 litres per minute. This rumen gas may have a carbon dioxide concentration as high as 80% some of which will be absorbed within the lung with resultant increase in arterial carbon dioxide levels (Patterson et al., 1965). This has an effect upon breathing patterns and respiration rates that are found to vary between 10 and 45 times per minute in resting giraffes.

Long necks contain long tracheas and although the flow of air in the trachea should be non-turbulent and laminar, there will be some frictional resistance. Estimations of the volume by determining water capacity give a volume of 2.5 litres in a 4.9 m animal (Patterson et al., 1965). Application of the formula for

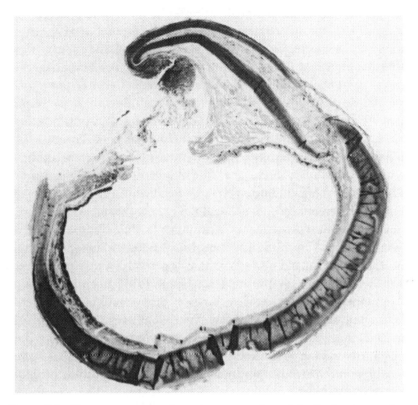

Figure 5.2. Transverse section through upper trachea of giraffe showing overlapping of cartilaginous ring to allow expansion.

calculating the volume of a cylinder: $V = \pi r^2 L$, where V = the volume in ml, r the mean radius in cm and L the length in cms, gives a volume of 1.6 litre. The giraffe trachea is in fact oval in section being capable of expansion and contraction (Figure 5.2). Harrison (1980) calculated trachea volumes in several specimens; a 15-year-old male had a trachea 2.3 m long with a maximum width of 51 mm reducing to 35 mm at its lowest margin. This gives an approximate volume of 3.0 litres although this is an underestimate of total dead space since it does not include the volume of the pharynx or bronchi. Similar measurements in a 13-year-old female gave a volume of 1.8 litres, neither correlated with dead body weight. By comparison with humans, the adult male giraffe is about eight times heavier. The volume of dead space in the latter is approximately nine times and total lung capacity eight times greater. Maintenance of adequate respiration in the giraffe employs two mechanisms to offset the large dead space. Economical respiration through a long tube requires slow infrequent breathing before lung ventilation can occur. In captivity, giraffes breath 8 to 12 times per minute unless disturbed. During exercise, and the giraffe can run fast when necessary, expiration may be assisted by its relatively large lungs and the ability to diminish the tracheal lumen by contracture of the muscle found posterior to its overlapping rings. A similar muscle is found in the horse and may serve a similar purpose. These physiological problems which are inherent in the respiratory system of the giraffe (and possibly other large herbivores) may play some part in the virtual absence of sound production in this animal. Low flow rates together with a large dead space may prevent vocal intensity reaching noticeable levels.

Morphological studies on the structure and development of the mammalian lung suggest that although lung volume for small and large mammals make up nearly the same percentage of body weight, there may be important quantitative differences between species (Tenny & Remmers, 1963; Maina, 1987) (Figure 5.3). The equation: $V_t = 53.5\ M_b^{1.06}$ suggests that lung volume relative to body mass is nearly constant, although there may be a trend towards an increase with body size in those mammals investigated by Stahl (1967). Schmidt-Nielsen *et al.* (1980) examined the structure of the lungs of shrews and found the density of gas-exchanging surfaces to be $1830\ cm^2\ cm^{-3}$ compared with $775\ cm^2\ cm^{-3}$ in the rat. In all other respects structural patterns were similar to those found in other mammals and the design and structure of the mammalian lung appears to reflect oxygen and metabolic demands which themselves can be attributed to body size and life-style.

Structural characteristics in the lungs of many terrestrial mammals have been studied by Gehr (1981), in bats (Maina & King, 1982) and in the baboon (Maina, 1987). Morphometric data give volume density of lung parenchyma as varying from 80 to 90% in mammalian lungs. In the horse is it 86%, wildebeest 87% and between 83 and 85% in six species of bat. In humans the volume density of 86.5% was close to that found in the baboon, 87.3%. However, Maina (1987) comparing human pulmonary parameters with a non-human primate (Olive baboon) and Thomson's gazelle (*Gazelle thomsonii*), concluded that the human lung was less well adapted for gaseous exchange compared with the lungs of more agile mammals. This will be considered in more detail in section 2.

Opening and closing the larynx

The limiting factor in flow of oxygen to the lungs lies at the level of the larynx which Fink (1978) called the 'modulated bottleneck'. Opening and closing is carried out by both intrinsic and extrinsic muscles although possibly utilizing

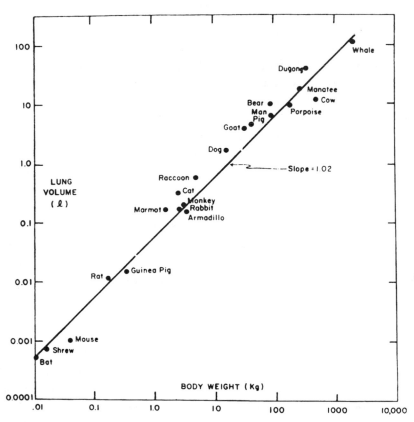

Figure 5.3. Logarithmic plot of lung volume as a function of body size. (After Tenny & Remmers, 1963.)

two separate mechanisms. Acting on the arytenoids, the posterior cricoarytenoid and synergistically the cricothyroid muscles abduct the vocal cords. Closure is by the lateral thyroarytenoid and cricothyroid muscles (acting without the posterior cricoarytenoid). These muscles act on the laryngeal airway by tensing the vocal cords or by movement of the arytenoid cartilage. Fink & Demarest (1978) postulated an additional mechanism which they called 'respiratory folding or bellows folding'. The laryngeal framework is suspended by muscles and ligaments from the hyoid bone and ultimately from the mandible and skull base. Attached to the larynx inferiorly are the trachea, bronchi and lungs, as well as several muscles. The larynx can therefore be considered as a system of suspended articulated cartilages lined internally by folds whose collective shape is altered by muscles, and restored by intrinsic ligaments (Fink & Demarest, 1978). Mechanically, ligaments attached to hyoid and thyroid transfer to cricoid and thyroid cartilages stresses generated by the vertical respiratory excursions of the larynx. Inspiratory enlargement of the larynx would involve the whole of the supracricoid larynx by pulling the hyoid and thyroid cartilages apart, that is, opening the bellows. Descent of the trachea will also increase the longitudinal tension in its walls stretching the elastic suspensory laryngeal ligaments. Together, these inspiratory efforts result in an 'unfolding' of the internal laryngeal surfaces, which is reversed by elastic recoil on expiration.

Konrad *et al.* (1984) measured spontaneous electromyographic activity of the laryngeal muscles simultaneously with respiration and airway resistance in 150 dogs. They verified the effects of muscle activity on airway resistance by denervating or stimulating muscles individually, and in synergic or antagonistic groups. They concluded that opening and closing mechanisms within the larynx were complicated, and agreed with Fink & Demarest (1978) that they depended on passive factors such as the mass and elastic forces acting through tongue base pharynx, larynx and trachea. Intrinsic muscles acting on the vocal cords were an essential component assisted by a 'bellows mechanism'; this appears to apply to all mammalian larynges.

Neurophysiology of respiratory function

Although respiratory activity in the cricothyroid muscle has been detected experimentally its functional significance remains unclear. This muscle contracts phasically during inspiration and stimuli such as airway occlusion or hypercapnia increase inspiratory activity. Muscle behaviour varies, with level of consciousness being minimal in alert rats and conscious humans. Woodson *et al.* (1989) measured laryngeal resistance in anaesthetized dogs together with the activity in the posterior cricoarytenoid and cricothyroid muscles. Paralysis of the latter had no effect on laryngeal resistance or the glottic area, even at high levels of activity. Spontaneous inspiratory cricothyroid contraction did not significantly affect laryngeal resistance calculated as the ratio between subglottic pressure and the constant airflow through the larynx. Artificially induced cricothyroid contraction by electrical stimulation of the external division of the superior laryngeal nerve increased laryngeal resistance, but Woodson *et al.* (1989) considered that physiological levels of respiratory activity in the cricothyroid muscle of the dog has no significant effect.

These findings are of clinical importance as it has been suggested that spontaneous contractions of the cricothyroid muscle are capable of adducting the vocal cords in patients with laryngeal paralysis (Dedo, 1970).

Airway resistance

To function efficiently as protector and controller of the lower respiratory airway, the larynx must achieve near perfect synchrony of its complex sensory and motor capabilities. The vocal cords act passively to restrict entrance of air into the trachea and consequently phasic contraction of the sole abductor; the posterior cricoarytenoid muscle is an essential part of successful respiration. Sasaki *et al.* (1973) showed this to be synchronous with inspiration and that the degree of abductor activity varied directly with ventilatory resistance. Resistance properties of the larynx are difficult to evaluate separately from the resistance of the tracheobronchial airway. However, its ability to undergo rapid adjustment corresponding to the phase of respiration and the physiological state of the lungs is recognized as being essential to all mammalian activity. The role of the cricothyroid muscle during inspiration remains in doubt but the posterior cricoarytenoid is clearly 'driven' by the medullary respiratory centre and the level of activity regulated by afferent impulses iniated by receptors within the larynx and lungs. Posterior cricoarytenoid muscle contraction increases the horizontal diameter of the glottis and it has been suggested that the anteroposterior length is increased by contraction of the cricothyroid, although this muscle appears to have a more important role during expiration (Sasaki & Buckwalter, 1981).

The resistance of the laryngeal airway constitutes an appreciable fraction of the total airway resistance in all mammals. McCaffrey & Kern (1980a,b) found that it varied considerably between species but remained fairly constant within individual animals. In spontaneously breathing dogs the laryngeal resistance during inspiration was approximately half that measured during expiration. Laryngeal resistance was measured by the pressure difference between subglottis and mouth while a constant flow of humidified air was passed through the larynx. Denervation of the larynx produced a rise in resistance during both inspiration and expiration, confirming that vocal cord abductors were tonically active during both phases of respiration.

Laryngeal resistance decreased in response to hypoxaemia although this proved less potent than hypercapnia. Such stimulation of chemoreceptors, whether physiological or pharmacological is accompanied by an increase in rate and depth of ventilation. McCaffrey & Kern (1980a,b) concluded from their experiments on the dog that reduction in laryngeal resistance was a direct reflex action of chemoreceptor stimulation, with a secondary effect through augmentation in tidal volume, stimulating stretch receptors in the lung. They found three types of pulmonary receptors: stretch, irritant and unmyelinated or J-receptors, in the human, cat, rabbit and dog. Each type evokes a specific reflex pattern of response in rate and depth of respiration, heart rate and blood pressure.

Stretch receptors stimulated by lung inflation inhibited phasic inspiratory activity in the laryngeal abductor muscles reducing laryngeal resistance.

J-receptors are physiologically located in that region of the lung supplied by the pulmonary circulation. This is consistent with the juxta-alveolar capillaries, hence the name J-receptor (Paintal, 1973). Stimulation produces an increase in laryngeal resistance during the resultant apnoea, a similar effect can be produced in the cat by phenyldiguanide (Stransky *et al.*, 1973).

Irritant receptors produce tachypnea with decrease in tidal volume in response to irritation, histamine and lung deflation. This is associated with a decrease in laryngeal resistance and serves to expel airway irritants.

Receptors of three types are present within the larynx. Flow receptors are stimulated by airflow, but in reality are cold thermoreceptors inactive at normal mucosal temperatures. Pressure receptors are the most common responding to changes in transluminal pressures, whilst drive receptors are physically stimulated during breathing (Petcu & Sasaki, 1991).

All receptors play some part in the adjustment of expiratory airflow and thereby lung volume in response to variations in laryngeal airway resistance. Rapid and reversible modifications of laryngeal resistance in response to respiratory needs of all mammals appear to be governed by this network of reflex arcs. However, the relative importance of individual receptors for specific needs, such as running, diving and other physical demands, remains unclear.

5.2 Locomotion

Measuring the role of the larynx in locomotion

The conception that structural elements are naturally formulated in such a manner that they satisfy, but do not exceed functional requirements, is termed symmorphosis. This is achieved by regulated morphogenesis and can be paraphrased as 'structure related to function'. Gans (1979) expressing a similar concept says 'most aspects of phenotypes will at any given moment in an individual's life be capable of fulfilling demands far greater than those routinely encountered'.

Mammalian locomotion, whether it is running, swimming, flying or some similar activity requires energy. It is logical therefore, to expect the larynx to play a significant role in all forms of locomotion in view of its control over the oxidative pathway. Although uneconomical for any mammal to possess potential capacity that is not to be used, it is maximal rather than basal metabolic rate that is most relevant to natural survival. The mitochondria in the muscles of a human runner can consume more than 5 litres of oxygen per minute. This requires a continuous flow of air through the larynx to the lungs, transference of oxygen to erythrocytes, circulation to appropriate tissues and finally diffusion to mitochondria in the muscles. Conversely, carbon dioxide produced during metabolism must be removed along similar pathways. Although subtle modifications of this process exist to meet specific needs within individual species, symmorphosis suggests that the structural design is optimized, controls the rate-limiting factor for oxygen flow at each level but is adaptable within certain limits (Taylor & Weibel, 1981). The basic tool for comparative studies is allometry or the examination of scaling of structures and functions relative to body mass (Huxley, 1932). Allometry literally means 'of other or different measures', and is used to describe differences in proportions correlated with changes in magnitude of the whole organism, that is, the study of size and its significance. Comparative studies have been carried out scaling organ weights with body mass, although the latter must be clearly defined with respect to whether the weight is of dead or living animals. Failure to differentiate between the two introduces an unquantified error into subsequent calculations. Sick or injured animals often have abnormal organs and in such instances inaccurate data will skew subsequent statistics. Allometric equations are no more than empirical descriptions relating some quantitative aspect of a mammal's physiology or morphology to its body mass. Underlying this is a fundamental similarity in biochemical pathways and morphological structure, for mammals are more similar to each other than they are to other species. Variations between species are largely due to disparity in size or natural behaviour with scaling best expressed by the equation:

$$Y = aM^b$$

Y is any physiological, morphological or ecological variable that appears to be correlated with size. The exponent b is the scaling factor, since it describes the effect of changes in body size, whilst coefficient a incorporates all dimensions required for consistency in this equation (Calder, 1984).

Experimental evidence correlates resting oxygen consumption rates with allometric expressions that might be expected to account for that consumption. In eutherian mammals lung mass in gram scales at $11.3 M^{0.99}$, lung volume in ml at $53.5 M^{1.06}$ and tracheal volume in ml at $0.82 M^{1.18}$ (Hinds & Calder, 1971).

Lung volume appears to represent about 0.6% and heart mass 1.15 of body mass, so that the work rates for respiration and circulation are a constant function of the metabolic needs, whether at rest or during sustained exertion. The design of the mammalian respiratory system appears to be devised for the maximum demands that will be required for survival. The alveolae interface has to be large enough to meet the oxygen requirements of an animal under maximum exertion. The structural design must include adequate circulatory, haematological and muscular facilities to satisfy all energy needs. Thus, maximum oxygen consumption should scale with body mass so that it bears a constant relation to the basal metabolic rate, that is with an exponent of 0.75. Seeherman *et al.* (1981) measured maximum oxygen consumption during treadmill exercises on mammals ranging in body mass from 7 g (pygmy mice) to more than 100 kg (ponies). When these animals ran at speeds exceeding maximal oxygen consumption they stopped as soon as blood lactate levels reached 18–28 mmol kg^{-1}. Taylor & Weibel (1987) using a similar treadmill technique studied the scaling of maximum rate of oxygen consumption for 14 wild and 8 domestic bovids. The slope for the wild mammals was 0.79 and domestic mammals 0.76. Combined data from Seeherman *et al.* (1981) and Taylor & Weibel (1987) show a maximum rate of oxygen consumption close to a constant multiple of 10 times the resting metabolic rate, scaling proportionally as $M_b{}^{0.75}$. This implies that, with increasing body size there is a commensurate increase in respiratory capacity, greater than might be expected for maximum aerobic capacity.

Locomotion whether for speed, endurance or simply foraging is the ability which characterizes all cursorial vertebrates. Although predators exploit superior speed, relay tactics, relentless endurance or surprise to overcome their prey, the latter frequently posses equal abilities, unless immature or infirm. The many factors inherent in natural survival will be discussed under specific locomotor activities but it is apparent that increased tidal volume is not the only way in which more oxygen is made available to muscles during increased activity. Respiratory studies on captive wild animals suggests that flow of air to the lungs is the single most important limiting factor in attaining maximum running speed or other escape activities. In the wild, a survival factor that relates to maximum aerobic performance achieves a balance reached between predator and prey. In comparing the morphology of the larynx and trachea in a variety of mammals, Harrison & Denny (1985) emphasized the importance of grouping data into similar behaviour groups. Zoological orders are based upon morphological characteristics which frequently bear little relationship to behaviour patterns or life-styles, for example, both the weasel and polar bear are carnivores but pursue very different lives. Even so, analysis of their data showed little relationship between glottic and tracheal size or body weight and maximum running speed. Although measurements of the glottic and tracheal area can be formalized by using clearly defined anatomical parameters, potential errors are present when using dead rather than live body weights. Data for maximum running speeds or other escape activities are unknown for many mammals or are based on experimental rather than natural timing.

If the concept of symmorphosis is accepted, then a significant relationship should exist between maximum glottic area (site of greatest airway resistance) and tracheal area, for most mammals. The phenomenon of allometry, defined as the study of proportional changes correlated with variation in size, should also exist when these areas are related to body size. Harrison & Denny (1989) repeated their measurements of glottic and tracheal area on a further group of mammals after removing all muscles attached to the arytenoids, whilst preserving the cricoarytenoid joints. In small specimens this was carried out under magnification. Thread was passed through the body of the arytenoid above the muscular process. By carefully applying tension the maximum glottic aperture was produced, the direction of pull varying depending on the size and angle of the vocal cords. Abduction was limited by the resistance of the mucous membrane, stiffness of cricoarytenoid joints and the effects of fixation. Where necessary, supraglottic structures were removed to facilitate access for photography (Figures 5.4 and 5.5). Scaled photographs of the glottic area and first tracheal ring were used for measurements utilizing a computer with graphic tablet. Data relating to age, sex, body weight and species were available for most specimens. The data were analysed using STATGRAPHIC software run on an Apple IIcx computer employing least squares regression analysis. The gradient intercept and Pearson's correlation coefficient were obtained for $\log_{10}$ scaled data; T statistics were also available and scatterplots allowed identification of 'outriders'. Potential errors in such an analysis are well recognized, particularly for biologically derived data. Significant regression, however, may suggest a dependent relationship between two variables, although influenced by the values chosen as 'dependent'. Presentation of the standard error of both slope and intercept is therefore essential when presenting final conclusions.

Tables 5.1 to 5.3 present the analysis of body weight to glottic area, body weight to tracheal area and glottic to tracheal area. Only a small number of points outside the 95% confidence lines were discarded, and variations in the number of observations included for analysis were due to incompleteness of relevant data in some specimens.

Table 5.1. *Regression analysis – dependent variable body mass, independent variable glottic area*

Parameter	Estimate	S.E.	T value	Prob. level
Intercept	1.92352	0.48165	5.48165	1.804E−7
Slope	0.52608	0.089022	6.39645	2.752E−9

n = 148
Correlation coefficient: 0.4644; S.E. estimation: 1.6358; R^2 = 21.57%.

Table 5.2. *Regression analysis – dependent variable body mass, independent variable tracheal area*

Parameter	Estimate	S.E.	T value	Prob. level
Intercept	−1.78728	0.278895	−6.41998	7.613E−10
Slope	1.01149	0.04762	21.2406	0

n = 235
Correlation coefficient: 0.8126; S.E. estimation: 0.9489; R^2 = 66.04%.

Table 5.3. *Regression analysis – dependent variable glottic mass, independent variable tracheal area*

Parameter	Estimate	S.E.	T value	Prob. level
Intercept	−0.20594	0.143526	−1.49069	0.1589
Slope	0.93395	0.02459	35.1091	0

n = 90
Correlation coefficient: 0.9689; S.E. estimation: 0.3935; R^2 = 93.7%.

Interpretation

The likelihood that a correlation is spurious or occurs by chance and recognition that sampling errors are inherent in all regression slopes, intercepts and correlation coefficients, restricts any conclusions drawn from this analysis. Input errors are not eliminated by statistical manipulation, and the choice of the 'dependent' variable may not always be obvious. However, correlation analysis is useful in estimating the extent to which two variables appear to show an association. Regression analysis is also useful in estimating the closeness of such a relationship. Good statistics are, however, no more than descriptions of

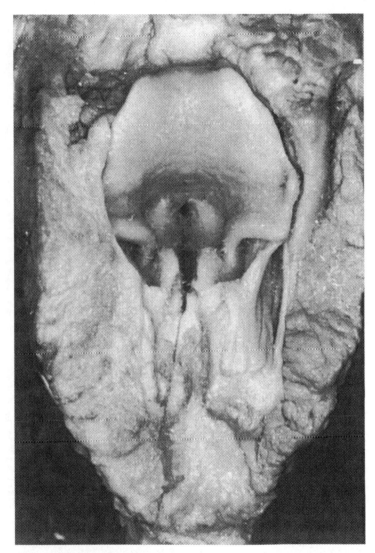

Figure 5.4. View of the supraglottic region of giant anteater (*Myrmecophaga tridactyla*) showing high aryepiglottic folds preventing a view of vocal cords. Note the presence of numerous fine hairs on the mucosal surface, a unique figure of this species.

available data. Even when judged as mathematically significant they may not indicate biological significance. By measuring maximal anatomical glottic space rather than the area produced by physiological abduction of the vocal cords, a difference of almost 30% was found in some specimens which had been

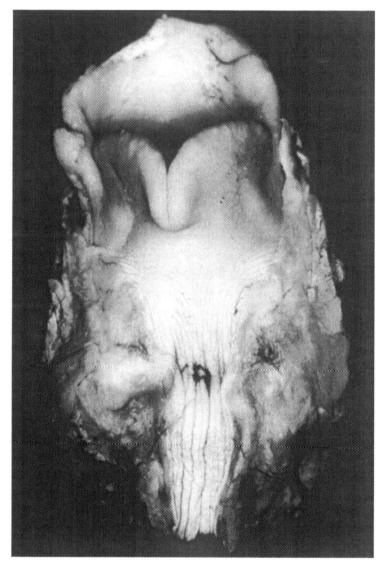

Figure 5.5. Macroscopical view of the larynx of the onager (*Equus hemionus*) showing the large arytenoids and high aryepiglottic folds typical of a herbivore. These prevent a clear view of the vocal cords.

preserved for many months. These data, however, represent the theoretical maximum space available within the larynx for airflow and is unlikely to be available under normal conditions. In some species such as the horse and giraffe, the tracheal rings are capable of physiological expansion and the area measured within the first ring is again a morphological compromise. Table 5.1 shows an exponent of 0.526 with confidence limits covering 0.609 and 0.483. Isometric scaling suggests that glottic area should scale as $M_b^{0.6}$ but this coefficient is low and there was some scattering of measurements. From this evidence it is doubtful if maximal glottic area can be considered as scaling closely with body mass. A closer relationship was found in the data in Table 5.2 where a slope of 1.011 (similar to that found between lung volume and body mass) suggests that tracheal dead space is a constant function of lung volume. The results of regression analysis for tracheal area against maximum glottic area suggest linear scaling. Correlation was high and the scattergram showed a good fit of data points within the 95% confidence lines.

The purpose of the original investigations of Harrison & Denny (1985) was to determine whether there was a direct relationship between glottic and tracheal areas with maximum running speed in animals grouped by natural behaviour patterns. As will be discussed under 'Running', this was restricted by lack of accurate data for maximum running speeds under natural conditions for many of the mammals examined. Data from Harrison & Denny (1989) appear to confirm that symmorphosis is present within the laryngotracheal complex and that the total respiratory complex is designed to provide for each individual's need for survival, under natural conditions.

Energy requirements for locomotion (Altmann, 1987)

With few exceptions food is the prime source of energy for all mammals and increases in energy expenditure must be met, eventually, by a greater intake of food. It is generally agreed that terrestrial mammals use less than 2% of their total energy expended on all activities, for locomotion. This is nearly independent of body size. Net cost of transport has been defined by Taylor (1978) as the metabolic power per unit mass during movement, minus the metabolic power per unit mass at rest, all divided by velocity. However, animals move not just one kilogram of body mass but the whole body, and must pay the total energy cost which increases in proportion to body mass to the power of 0.7 (Altmann, 1987). There is, however, variability in published regressions of daily expenditure against body mass for most free-ranging wild animals. Available data suggest that the unit cost of locomotion appears to be little affected by body mass, and in most mammals is relatively small compared with total energy requirements. The upper limit of sustained activity appears to be controlled by the rate at which mammals can carry out locomotion aerobically, without accumulating unacceptable levels of lactic acid. If size is a locomotor advantage then being large may not lead only to conservation of energy but allow more time to forage. By covering greater distances in a given time, large animals increase their opportunity for finding food; although distance and time have differing size-dependent significances for individual mammals. Mammals occupy a wide variety of habitats requiring free movement on land, in water, in air and in the trees. Each activity demands adequate supplies of oxygen and the following sections discuss the role played by the larynx and respiratory tract in meeting these requirements.

Running

Survival for most terrestrial mammals is dependent on a finely balanced relationship between prey and predator. On the open plains lack of speed for potential predators would be a vital factor in subsequent starvation. For potential prey in a similar environment, the best defence may be exposure with ability to detect and escape from danger by high speed evasion. Although the cost of locomotion for foraging or hunting is therefore an important facet of physiological behaviour, survival for most terrestrial mammals is related to acceleration and maximum running speed.

If the premise is accepted that the larynx represents the most important area of restriction within the respiratory airway, excluding variations and complexity in shape and dimensions of the nasal airway, then when maximal aerobic capacity is reached the limiting factor to pulmonary aeration could be the maximum glottic area. This is oversimplification of the physiological state because other factors such as cardiac output, muscle action, limb morphology, etc., will play a part in individual species locomotor performance. Quadrupeds share muscles participating in both locomotion and respiration, resulting in a locomotor respiratory coupling between ventilatory and stride frequency (Bramble & Carrier, 1983). Mammals of different sizes change gaits at different speeds and stride frequency and length play essential roles in determining maximum speed. To maintain a continued linear increase in oxygen uptake with increased speed, the animal must take deeper breaths within the same time period. Yamashivo & Grodin (1973) concluded that as a consequence of the elastic resistance of the thorax, airway resistance and dead space, there is a specific rate of breathing for any minute volume that maximizes ventilatory efficiency. This would result in an optimal running speed for each gait which Hoyt & Taylor (1981) found in the horse.

Measurements of maximum aerobic capacity take no account of optimal speeds or maximum efficiencies. The accessory muscles of respiration in forcing the thoracic cage and lungs to oscillate faster when running are effectively resisting the impedance of the respiratory system. If this is limited by the muscles' physical capability then the only means of increasing ventilation would be to increase tidal volume. This will be ultimately determined by the maximum driving force and airflow impedance of which the resistance provided by the glottic aperture may be the most important factor.

Walking and hopping

Much research into the mechanics and energetics of locomotion has been carried out over the last two decades by workers such as Taylor (1978) and McNeill Alexander (1982). Bipeds and quadrupeds spend much of their time hopping or walking, changing gaits at different speeds, although at about the same Froude number. The latter, a term commonly used in nautical engineering, is defined for a running animal as u^2/gh, where u is the velocity, g the acceleration of free fall and h the hip height. Froude numbers are dimensionless so have no units, but are useful when comparing locomotion. Most bipeds run (like humans) or hop like kangaroos. At speeds below six km per hour (about 3.7 miles per hour) kangaroos do not hop but move slowly with their heavy tail as support. This gait has been called pentapedal (the tail acting as a fifth leg) and is clumsy, although hopping is costly in energy requirements. Shifting to quadrupedal locomotion at low speeds, as small hopping animals do, is precluded in kangaroos by the specialization of their hind limbs and reduction in size of their forelimbs. At low speeds the hopping gait has marked advantages to running, for the energy costs do not vary over a wide range of speeds. At speeds above 15 km per hour they hop more economically than running quadrupeds of a similar size, with their most economical speeds being between 22 and 25 km per hour. Above this speed energy costs increase because elastic storage of energy in their leg tendons cannot increase indefinitely (McNeill Alexander, 1988). Kangaroos can increase their speed above 40 km per hour in high speed bursts for several hundred metres in an emergency. There is then a noticeable increase in both stride length and frequency (Dawson & Taylor, 1973).

Marsupials have been considered as exhibiting lower or more primitive features such as an external pouch, epipubic bone and a lower metabolic rate than placentals (Dawson, 1983). Hopping uses much energy and with a low metabolic rate marsupials may have a diminished potential for speedy locomotion. Economy in long-distance travel, however, may be as important for some mammals as maximum sustained speed. In metabolically limited animals such as the kangaroo and other macropodids, bipedal hopping may be of greater economy in long-distance foraging than a high 'escape' speed. Garland *et al.* (1988) compared the locomotor performance of a group of marsupials with data relating to placental mammals. They concluded that the two groups showed no general difference with regard to maximal running speeds with interspecific differences being related to variations in body size, habitat and life-style. Maximal running speeds and maximal aerobic speeds (measured on a treadmill) were similar in both marsupials and placentals.

Bennett (1987) studied five species of Macropodidae ranging from potoroo (weighing 0.5 kg) to kangaroo (60 kg) pursuing them in their natural environment. The kangaroo changed from walking to hopping at Froude numbers around 0.6, similar to the number at which quadrupedal mammals change from walking to a faster gait. Maximum observed speed was found to be proportional to body mass, unlike that found by McNeill Alexander *et al.* (1977) in African ungulates.

Locomotor techniques

Horses walk slowly, quicken to trot, canter and finally gallop at their highest speeds. Varying gaits involve different patterns of leg movements (McNeill Alexander, 1982). Before considering some of the factors which may influence the locomotor performance of individual mammals some consideration must be given to the association of gait with aerobic respiration.

Although bipeds either hop or run, quadrupeds use a wide range of running gaits, changing at different speeds but generally at the same Froude number. Horses trot at low speeds, changing gaits with increases in speed, finally galloping. Camels, on the other hand, are unique being the only species which invariably pace, i.e., the foreleg and hind leg on one side move forward simultaneously alternating with those on the other side. Other mammals that can pace are horses, dogs and bears, although they normally trot at low speeds. Camels use pace instead of trotting possibly because of their need for prolonged endurance rather than speed in the hot, arid regions in which they live. They can travel 150 km in 15–20 hours or 800 km in 11 days, all at a steady speed. The legs and feet are not like other Artiodactyla, for they have no hooves but large pads with two anterior toe nails and long legs. This is ideal for soft sand but disastrous in mud or on rough hard ground (Gauthier-Pilters & Dagg, 1981).

Large mammals such as the elephant and polar bear generally amble rather than trot and although having no really fast gait can travel embarrassingly quickly. The relative phases of the amble are similar to the quadrupedal walk, but the polar bear has been found to use oxygen at twice the rate expected for their size and speed of movement. 'Galloping' is probably only used in an emergency and then for no more than a minute or two because of the danger of heat exhaustion (Pond, 1989). Other mammals with thick fur such as reindeer (*Rangifer tarandus*) experience the same problem; even in subzero temperatures they may die after only several metres at full gallop. Other factors which have a restricting influence on gait are body weight and leg length. Elephants can manage only 5 m per second if the stresses on their relatively long legs are to be kept within acceptable limits in their straight-legged gait. By comparison, the rhinoceros has short thick legs with bones strong enough to achieve speeds up to 7.5 m per second. McNeill Alexander & Pond (1992) investigated stresses in the legs of these large mammals concluding that the rhinoceros skeleton may be built to a particularly high level of safety which makes it such a formidable adversary at speed.

Humans break into a run and kangaroos change from walking to hopping at about the same Froude number (0.6) at which quadrupedal mammals change from walking to a faster gait. There is, however, evidence suggesting that for each individual gait there is an optimal speed. At high speed most gaits are equally economical except for the pronk or stott. Faced with a predator such as the cheetah, many deer or antelope will bounce up in the air keeping all legs straight. Not only is this an expensive use of energy but makes the animal visible, and vulnerable to the predator. Prey normally stott when fleeing but it is normally carried out about 67 m from danger and is now thought to be a signal to the predator that it has been detected. Cheetah's usually need to start their run 20 m from their prey to have any chance of success. However, stotting and pronking are also known to occur in the absence of predators and this potentially wasteful gait must have other less obvious purposes.

McNeill Alexander & Jayes (1983) proposed a theory of dynamic similarity of gait. 'Different mammals move in a dynamically similar fashion whenever they travel at speeds with equal Froude numbers.' For cursorial mammals changes in speed of gait, stride lengths and performance can all be predicted for differing species with confidence. However, once running begins there is little increase in stride frequency in the larger animals; increased speed is attained by lengthening the stride. This correlates in most mammals with a linear increase in oxygen consumption up to the point of maximum aerobic capacity; it is this limiting factor which must be considered when comparing maximum running speeds

Maximum running speed or endurance

Speed and endurance characterize all cursorial vertebrates, that is those animals which can run far, fast and easily. It is difficult to find reliable data about the distances mammals travel during their normal activities or of their maximum running speed when hunting or threatened. Many herbivores feed in relatively restricted areas but wander long distances during seasonal changes. Wildebeest may travel more than 1000 km to take advantage of variations in the timing of the rainy season and availability of fresh grass. Large predators, capable of fast speeds over short distances stalk their prey so as to get as close as possible; whereas 'pack hunters' such as wolves or hunting dogs, tend to exhaust their prey by relentless pursuit over several kilometres. It has long been assumed that predators select old, weak or sick members of prey populations to reduce the risk of injury and also limiting energy expenditure. FitzGibbon & Fanshawe (1989) compared the age and condition of Thomson's gazelles (*Gazella thomsoni*) killed by wild dogs (coursing predators) and cheetahs (stalkers). As was expected, coursing animals with time to select weaker prey took more young, old and sick animals. Cheetahs, relying on surprise and short pursuit, have less time for prey selection and killed the most available animal. Similar conclusions have been drawn from studies of puma (*Felis concolor*) and wolf in North America. Such assumptions are however influenced by habitat, for dogs encountering prey in poor visibility or dense cover chase for short distances and take prey in good condition.

Measurements of maximum running speed made experimentally on treadmill or by 'vehicle chasing' have been carried out by McNeill Alexander *et al.* (1977), Garland (1983) and others for a variety of mammals. These techniques are restricted to larger animals or to a relatively open, reasonably flat environment. Consequently, the data are limited and in many instances questionable, although of value when correlated with observations of natural behaviour of individual species.

The basic physiology and mechanics of mammalian locomotion has been studied closely with particular emphasis on the analysis of the respiratory, circulatory and muscular modifications that are found in mammals of varying running capabilities (Brazier Howell, 1944; Jones, 1972; Day, 1981; McNeill Alexander, 1983). All exist for the primary purpose of survival and in fit animals, under natural conditions, it would be expected that neither prey nor predator had an overwhelming advantage. Most mammalian locomotion is carried out under aerobic conditions, except perhaps in the cheetah which may breath-hold for its short, high speed sprint. In their investigation of the influence of laryngeal and tracheal size on the running speeds of a variety of mammals, Harrison & Denny (1985) were unable to identify any general relationship that could be applied to all species. By grouping animals within similar behaviour patterns rather than maximum running speeds, it was anticipated that the degree of glottic 'choke' would be similar. This proved to be so for zebra, onager, horse and bison but surprisingly not for the pack hunters – wolf and wild dog. If maximal aerobic capacity has 'survival value' it is surprising that large carnivores should have relatively large glottic and tracheal areas, suitable for long distance locomotion. Antelopes who might be expected to have capability for long distance foraging appear to have a proportionally smaller maximum glottic–tracheal area for their body size. Such conclusions may be related to inadequate or unreliable data, but may also represent variations in energy efficiency.

Breeding and training for speed and endurance

For centuries man has attempted to improve the natural speed of horse and dog by breeding programmes, often carried out with scant knowledge of the underlying physiological processes inherent within these species. More recently attention has been paid to a less exotic animal, the camel. Thoroughbred camels are amongst the most prized possessions in many Middle Eastern countries, as well as being one of the most physiologically resilient of all mammals. These racing camels, bred from local wild animals, are small, fine-legged and female. Taking five years to reach maturity, selective breeding is hampered by long pregnancy (13 months) and poor natural reproductive efficiency, for their ovaries only release mature eggs after mating. Sudanese dromedaries can run over 10 km of sand in 18 minutes at an average speed of 33 km an hour aided by muscle fibres devoted to sustainable aerobic respiration, that can be maintained for the whole of an 18 minute race. There is no evidence that breeding has enhanced the morphological size of the respiratory pathway and performance appears to depend upon training, feeding and veterinary care (Vines, 1992).

Greyhounds can run faster than horses and although possessing excellent aerobic capacity have limited long distance endurance. Breeding has produced an animal with over 60% of body weight composed of 'fast twitching' muscle fibres. This results in rapid acceleration followed by early fatigue from lactic acid accumulation. As with thoroughbred horses, heart size is greater than in the naturally occurring dog. Gunn (1989) weighed the hearts of 11 thoroughbreds, 12 other horses, 18 greyhounds and 9 other dogs. The fast running animals all had larger hearts relative to live weight than their slower relatives. Although this is unrelated to variations in stride length and frequency, increased cardiac stroke volume is advantageous in increasing circulation time and therefore tissue metabolism (Vines, 1987). Measurements of the maximum glottic and tracheal area were carried out by Harrison (unpub. data) on 14 greyhounds and 20 'assorted' dogs of similar body weight, no statistical significant difference in laryngotracheal ratios were found. There is no reliable evidence to suggest that breeding does more than produce dogs with large amounts of 'racing muscle' who are fit and trained for competitive racing.

The racehorse

It is more than likely that a similar situation applies to the most popular of all breeding experiments, the thoroughbred racing horse. The main predator of the wild horse is a carnivore (usually a lioness) whose maximum speed is limited to 100 m. To survive, the horse has only to run a little faster for a little longer and it is unnatural to expect it to run a mile (*c*. 1.6 km) or more simply for survival. Flight is the horse's reaction to danger and its best defence. Although occasionally running for sheer exuberance or excitement the racing horse has to be encouraged to run distances that far exceed its natural requirement.

Measurements of oxygen consumption during endurance exercises indicate that at rest the thoroughbred uses between 2 and 5 ml kg^{-1} of body weight per minute rising to between 48 and 65 ml kg^{-1} min^{-1} on cross country events. During a gallop this can rise to 160 to 200 ml kg^{-1} min^{-1}. However, 80% of this energy is wasted as heat, and this poses considerable problems when racing in hot climates (Bonner, 1994).

A thoroughbred racing over the 'Classic' distances will have a respiratory rate of around 140 a minute representing two and a half breaths a second. Inhaling 10 litres of air at each breath amounts to a minute volume of 1400 litres and represents a major component of the total work in galloping (Cook, 1989). The horse larynx and trachea are more than adequate for such airflows, but Cook has produced evidence suggesting that 95% of thoroughbreds, whose average height is greater than '15 hands', have some restriction in vocal cord movement from unilateral recurrent laryngeal nerve neuropathy. Such limitation in air flow will affect racing performance, something recognized since the time of the Prophet Mohammed when the finest horses were those that had 'strength and wind'. Horses are still sold as being 'sound in limb and wind' (see section 4.6).

Cook (1989) believes that in 99% of instances nerve paresis is inherited, the left side being the most frequently affected. This may be one of the factors that is responsible for the failure of racehorses to run faster today than their counterparts 50 years ago. Inbreeding has existed since the establishment of the Stud Book in 1791 and it is estimated that more than 80% of the genes from today's thoroughbreds come from 31 horses alive in the eighteenth century (Dunbar, 1985). Breeders work on the principles that characteristics such as height at the shoulder, heart size and body length are all inherited and a large percentage of Classic winners come from sires or mares who have been successful in these races. Yet despite systematic breeding and improvements in racing surfaces there has been stasis in winning times. Although intensive inbreeding has not produced serious genetic side effects it appears that genetic limits have been reached for improving the physical abilities of racehorses. This is hardly surprising since the rationale behind breeding philosophy has never been based on biological criteria but rather on the assumption that 'Classic winners breed true'. With an 11-month gestation period and peak racing performance not reached until the age of three or four years, it is many years before a stallion's 'prowess' can be quantified. There is no evidence that thoroughbreds have significant improvements in laryngotracheal airflow over their wild relations. Racing success then becomes a matter of training, veterinary care and 'motivation', which Cook (1976) believes to be largely related to fully functioning vocal cords.

Human runners

Despite considerable personal effort and scientifically planned training, humans compare unfavourably with most mammals in athletic ability. Greyhounds and racehorses sprint faster and in most instances have greater long distance endurance. Even warthogs can run as fast as humans and the peak sustainable speed of the domestic pig is only slightly less than 6 m per second. The bipedal mode of walking by humans seems potentially catastrophic; only the rhythmic forward movement of first one leg and then the other keeps him or her from falling flat on his or her face. It is, however, a unique form of locomotion, for although most primates can stand on their hind limbs only humans can walk with a striding gait (Napier, 1973).

All humans have the ability to move from walking to running, although in the unfit, duration is limited for both both speed and endurance. As with other terrestrial mammals, human running speed is a product of striding rate and striding length, with the former pre-eminent in the best sprinters. Over a distance of 100 m run at 12 m per second (28.8 miles per hour) an athlete will take between 4.40 and 5.02 strides per second. Excessive striding rates deplete energy stores quickly and lead to a loss of speed in the last part of a race. Some coaches emphasize increasing stride length, although it is an individual's combination of stride length and stride speed which produces the best result. In every cursorial mammal stride length is composed of ground-contact time and airborne time. Only the former produces the force necessary for running; force platforms being used to measure the force exerted on the feet during various sporting activities. McNeill Alexander (1988) experimented on human feet removed for disease, concluding that a 70 kg man running at 4.5 m s^{-1} stores and regains about 100 J energy. About 35 J was stored as strain energy in the Achilles tendon and 17 J in the arch of the foot. Humans lack much of the elastic mechanisms which prove so effective in fleet-footed animals such as zebras, giraffes, horses and antelopes. On soft ground humans run well in bare feet using energy stored in their Achilles tendons. Modern running shoes attempt to return this energy to the runner's foot by incorporating carbon-fibre leaf springs storing energy on heel impact and restoring it to the toe region. All-weather running tracks are now designed to absorb between 35 and 50% of the foot's impact. Specifications dictate that the synthetic surface layer should not be thinner than 12 mm nor deform under foot pressure by less than 0.6 mm or more than 1.8 mm (Goldspink, 1992). Athletic training incorporates not only track coaching but often intensive evaluation by sports scientists utilizing their knowledge of biomechanics, physiology, nutrition and sports psychology.

Recent developments in exercise physiology emphasize the importance of genetic control in the adjustment of muscle size and behaviour in response to mechanical stimulation. In the mid-1970s, biochemical studies demonstrated three main types of muscle fibres: 'fast' fibres programmed for rapid contractions, 'slow' fibres for repetitive longer-lasting contractions and an intermediate type for relatively fast and long-lasting work. Muscles contain a variable amount of these three types depending on their function. Myosin plays an essential role in all muscle activity and mammals possess at least seven different versions of the gene that encodes the myosin heavy chain of skeletal muscles. A specific version of the myosin gene has been found in the jaw muscle of carnivores and is thought to enable jaw muscles to contract even faster then conventional 'fast fibres' (Goldspink, 1992).

Different exercises encourage different muscle fibres to enlarge and in sprinting it is the total output of power that is important. This is related to speed of muscle contraction and production of maximum force. Good sprinters will need a high proportion of fast fibres whilst long distance runners require fatigue resistant muscles. Animal experiments confirm that controlled exercise can influence selective growth of specific muscle fibres but not the total number which appears to be genetically determined. Quality athletes are born not made!

Despite the need for plentiful supplies of energy to their musculature there is no evidence of any marked increase in heart size of trained athletes. Normally, humans can increase their cardiac output to 15–20 litres per min during severe exertion; experienced athletes can sustain an output at almost double this level. Training does slow the resting heart rate but this is no predictor of success, and heart size of most experienced athletes falls within the range of the normal population. There are in fact no reliable data to support the concept that successful athletes differ morphologically from their less active compatriots. Indeed, although regular physical exercise is accepted as being beneficial in reducing the incidence of heart disease and hypertension there is evidence that intense activity may predispose to infections by depressing the immune system

(Sharp & Parry-Billings, 1992). Overtraining, resulting in prolonged fatigue and reduced performance from diminished aerobic capacity and muscle force, is well recognized. There is evidence to suggest that these symptoms may be associated with immune depression as measured by T-cell activity and research into this concept is progressing.

An increased need for carbohydrate is a major effect of intense training, because carbohydrate is the principal source of muscle energy. During slow walking the muscles mostly use fat in the form of free fatty acids. In fast running energy needs are met from carbohydrate and many athletes in training consume more than 600 g of carbohydrate daily (Turnbill, 1992). They require a high carbohydrate diet to compensate for this high energy expenditure and the dietitian has become another member of the 'athletic production team'.

A highly motivated human backed by knowledgeable specialists in sports medicine, exercise physiology and dietetics, coached by experts and running on the finest tracks in specially designed shoes can never match the 'natural' runners, such as cheetah, lion or horse. Training undoubtedly enhances natural muscular ability and probably leads to more effective oxygenation of the muscles. There is, however, no evidence that any individual athlete has a laryngotracheal airflow greater than that found within the normal 'untrained' population. Success therefore appears to be the result of a combination of intense dedication and enhancement of inherited physical attributes. A situation similar to that found in other mammals.

Climbing

Negus (1947) discussing the functional significance in mammals of a 'valvular' larynx associated it with independent use of the forelimbs, as in grasping, hugging, clinging or climbing. This can best be studied in primates. Martin (1990) categorized the various locomotor techniques found amongst living primates. This is based on earlier studies by Napier & Napier (1985) and divides groups into vertical-clinging and leaping, arboreal quadrupedalism, terrestrial quadrupedalism, arboreal arm-swinging and terrestrial bipedalism.

Vertical clingers predominately grasp vertical tree trunks before leaping between them, and usually have well developed hind limbs. Species include bushbabies (Galago sp.), lemur (Lepilemur sp.) and tarsiers (Tarsius spp.). Squirrels can travel in a similar manner but have claws which they dig into bark for added security and speed. As the name indicates, quadrupedal primates use both fore- and hind limbs together in a variety of differing sequences of locomotion. This is the basic pattern for most terrestrial primates such as the baboon, as well as many predominantly arboreal species of New and Old World monkeys. The dominant component of brachiation is emphasis of the forelimbs rather than exclusive use of arm-swinging as in gibbons (*Hylobates spp.*). This is now differentiated from the more deliberate suspensory locomotion found in great apes by calling the latter 'modified brachiation'. The great apes have long arms, hook-like hands and wide chests. Whilst the chimpanzee spends half of its waking life on the ground, the gorilla is almost completely terrestrial.

Classification of primate locomotor patterns is useful in comparative studies although most species utilize a variety of techniques. Rapid and often precarious movements made to escape danger may not be seen during normal behaviour but reflect the capacity of the species to survive. Similar criticism can be applied to an assumption that locomotor categorization matches taxonomic classification and offers no additional information. For example, small-bodied vertical-clinging groups include representatives of two widely separated taxonomic categories: Galaginae and Tarsiidae (Martin, 1990).

Morphological adaptations and evolutionary significance of the various forms of primate terrestrial and arboreal locomotion are complex and consideration of them is beyond the scope of this chapter. Climbing is part of the normal behaviour of many mammals and Negus (1947) found that all climbers had larynges capable of preventing ingress of air, closing the glottis and, thereby, fixing the chest muscles to allow the use of the pectoral muscles for climbing. Not every mammal with this facility climbs for it also allows grasping and hugging, but no mammal that climbs is without a valvular type larynx (Figures 5.6 and 5.7). Development of a valve in the larynx by inferior thyroarytenoid folds turned upwards is found in many mammals (Figure 5.8 and 5.9), particularly in active arboreal species such as lemurs, gibbons and marmosets. A secondary sphincter formed by 'downturned' false cords preventing exit of air is also seen in the great apes and some other species. This may be associated with an expansile air sac leading from the ventricle which lies between these two sphincters. Section 4.3 devoted to ventricles and saccules discusses the frequency and significance of air sacs. Negus (1947) believed that these were used for re-breathing during brachiation. This is unlikely, for both chimpanzee and gorilla spend most of their locomotor activity on the ground, yet both species have well developed air sacs. Another great ape, the orang-utan, an arboreal mammal, has a strikingly different gait. In trees it is quadrumanal, using both hind and forelimbs to suspend the body; movement is slow and circumspect with plenty of opportunity for respiration. If rebreathing from sacs which lay beneath large contracting pectoral muscles was feasible, then the inferior thyroepiglottic folds would have to be open whilst the superior thyroepiglottic folds were closed.

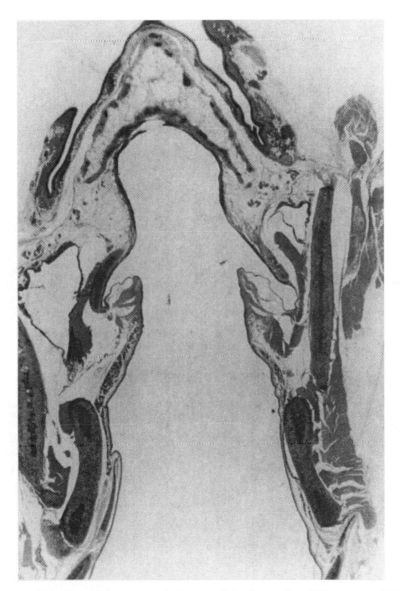

Figure 5.6. Coronal section of the larynx of the fennec fox (*Vulpes zerda*) which although not a climber, has a valvular larynx, upturned vocal cords and ventricles.

There is no evidence that this happens and the purpose of the laryngeal air sacs remains uncertain. Humans, whilst possessing both vocal cords and ventricular bands, have no air sacs and are poorly designed for climbing.

All forms of running, climbing and jumping use energy. Although analysis of

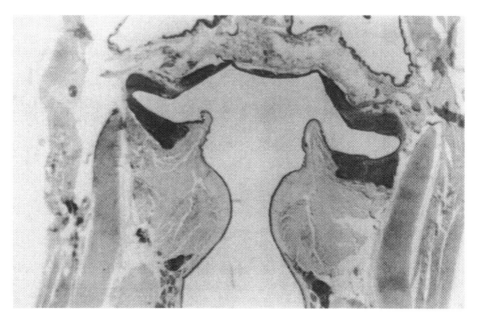

Figure 5.7. The domestic cat is an excellent climber having a valvular larynx but no ventricles or false cords.

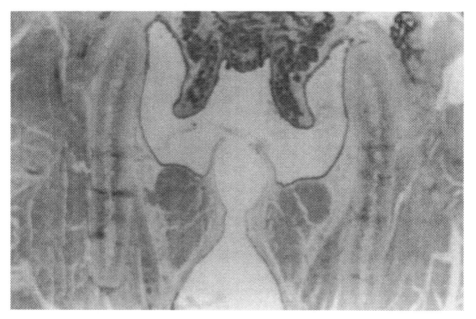

Figure 5.8. Thin, upturned vocal cords with a birfurcated epiglottis are found in the emperor tamarin (*Saguinus imperator*).

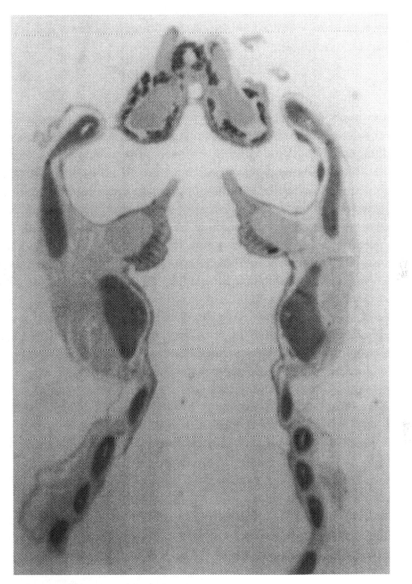

Figure 5.9. Thicker upturned vocal cords are found in the northern night monkey, (*Aotus trivirgatus*), another good climber.

the mechanics of each activity reveals complex variations in morphology, in every instance the larynx plays an essential role in control of airflow. In addition the presence of valvular vocal cords capable of preventing air entering the trachea allows independent usage of forelimbs. Retention of air within the lungs is possible in mammals with ventricular bands capable of tight closure. These are associated with ventricular air sacs in gibbons and the great apes, although their purpose remains conjectural.

Flying

The most important adaptation for any flying animal is a large enough wing or membrane to support its weight. In mammals there are only a limited number of possibilities and both flying squirrels and bats utilize membranes around body and limbs. Bat wing design leads to greater agility and is ideal for the foraging of insects but requires a higher consumption of energy than birds. Some wings are long and narrow, others short and broad depending on body weight and foraging needs. Bats also vary in size from about 3 to 1400 g and many size dependent factors influence wing design. They show the expected increase in loading with body mass of approximately one-third power (Lighthill, 1977). Wing morphology and wing beat kinematics are correlated with life-style and flight habits so that energy needs are minimized. The evolution of this highly successful and specialized mammal has been associated with major physiological and biochemical adaptations essential for providing wing musculature with adequate oxygen supplies. In small bats such as *Myotis* spp. all muscle fibres are fast-acting in association with rapid wing beats and a high metabolic rate. As body mass increases so does the proportion of slow-twitch fibres possibly reflecting lower wing-beats and metabolic rates in larger bats (Vaughan, 1985). The oxygen used in flight by some species of bats has been measured in wind-tunnels. Rates of oxygen uptake indicated metabolic powers of between 50 and 125 W kg^{-1} body weight (McNeill Alexander, 1983).

Various wing parameters such as span, area and loading, scale in different ways in different groups of bats. Plots of wing loading and aspect ratios versus body mass show similar slopes for frugivorous microbats and the large megabats, whilst the insectivorous families have similar but steeper slopes for both wing loading and aspect ratios (Day, 1981). All fruit eaters show close convergence in wing form and have similar diets and foraging patterns; it is the manner in which these species exploit their environment that influences laryngeal morphology.

Bat larynx
There are about 1000 species of bats in the world, the order Chiroptera being divided into two suborders: the Megachiroptera (composed largely of large fruit bats and flying foxes) and the Microchiroptera whose diet is insects, fish, frogs, etc. (Corbet & Hill, 1991). Gross laryngeal morphology has been considered

already (Chapter 4) where the large musculature and ossified cartilages were seen to be a feature of the bat larynx (Figures 3.6, 4.29 and 4.30).

Larynges from 18 species of bats in both suborders have been studied by serial sections, the findings being described by Denny (1976). Bats have unique respiratory problems, as they require large volumes of oxygen for rapid aerial activity and long distance foraging and many species need to echolocate whilst 'on the wing'. The Megachiroptera, which are largely frugivorous or nec-tarivorous do not echolocate, although the male hammer-headed fruit bat (*Hypsignathus monstrosus*) is known to use penetrating calls during mating.

Rousettus aegyptiacus, however, is said to use ultrasound whilst roosting in dark caves, but this is produced by tongue clicking rather than the larynx (Pye, 1968). Fruit bats, however, are both gregarious and noisy; their larynges have typical mammalian vocal cords but have an expansion of the airway in the region of the posterior commissure. This reduces the constricting effect of the vocal cords during 'flight vocalization', and permits adequate air flow to the lungs (Figures 5.10 and 5.11). Certain of the echolocating members of the suborder Micro-chiroptera possess a unique morphological feature in upper tracheal air sacs. These vary in size, some appearing to be expanded tracheal rings, whilst others outpouch from beneath the vocal cords. Species without air sacs have ventricles; no saccules were found in any of the specimens examined. The significance of these air containing cavities in relation to echolocation is discussed in the section 5.3.

Diving

Fully developed aquatic mammals have lungs and therefore need to breathe air. The physiological problems inherent in the need to breathe whilst living in water and particularly the need to dive, dictate that they must utilize a limited oxygen supply with maximum efficiency. With the exception of Chiroptera, Lagomorpha and Primates, aquatic animals are found in all the mammalian orders, with seals, whales/dolphins and sea cows being fully aquatic. The manner in which these mammals overcome the physiological demands of oxygen conservation appear to be similar in both diving and non-diving species, although varying in degree. These have been discussed in detail by Elsner & Gooden (1983) and Schmidt-Nielsen (1983), and are best illustrated by fully adapted species. The most successful diver amongst the Pinnipedia is the Weddell seal (*Leptonychotes weddelli*) which has been recorded as diving to more than 600 m, remaining submerged for over one hour. Their lungs are small relative to the volume of the trachea; increasing hydrostatic pressure collapses the alveoli forcing air from lung to tracheal dead space and reducing gaseous exchange and nitrogen absorption. Partial pressure of blood nitrogen measured in these seals at depths of 230 m shows peaking at 70 m but continues to fall during further descent. Seals exhale before deep diving, limiting the amount of

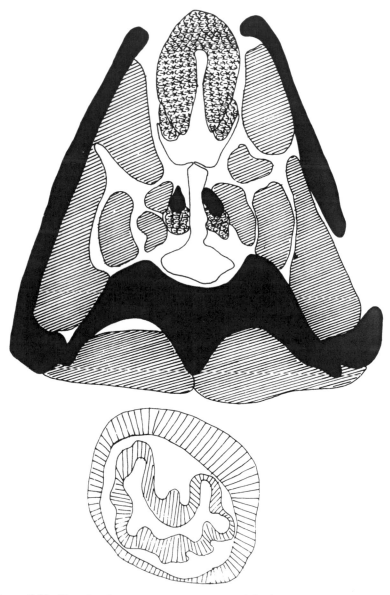

Figure 5.10. Drawing from a transverse section of the larynx of sac-winged bat (*Saccopteryx bilineata*) showing posterior 'air' gap.

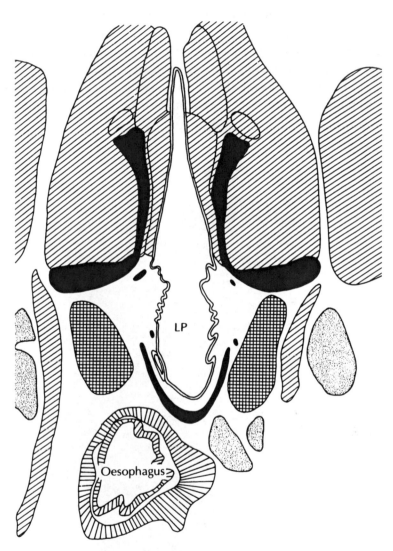

Figure 5.11. Transverse section of the larynx of the moustached bat (*Pteronotus parnellii*) showing posterior expansile area (LP) which may also be used in the production of ultrasound.

nitrogen in their lungs and the amount available for resorption on ascent. Whales, however, submerge without exhalation although it has been calculated their relatively small lungs are completely compressed at 100 m, forcing air into the dead space. If the effect of cold is ignored then the duration of dives is related to the speed of oxygen depletion.

Oxygen carrying capacity of most of the best animal divers ranges between 30 and 40 ml per 100 ml blood which exceeds that of terrestrial mammals. Whales have high concentrations of myoglobin in their skeletal muscles and this appears to be responsible for increasing their oxygen carrying capacity. An additional factor found in all diving mammals is pronounced bradycardia on submersion, this being observed not only in cetaceans but also muskrat (*Odontara zibethicus*),

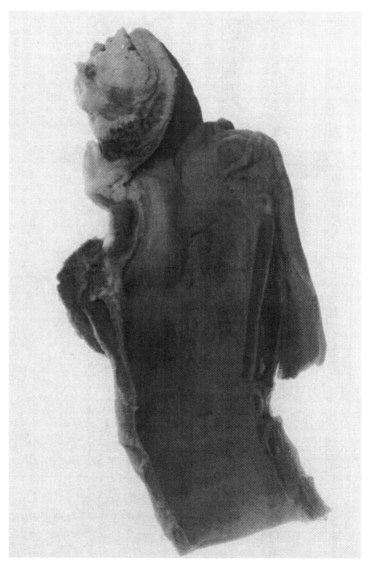

Figure 5.12. Section through the larynx and trachea of the grey seal (*Halichoerus grypus*) showing the large trachea in relation to the relatively small glottis.

hippopotamus and humans. Despite these physiological modifications it is doubtful if adequate oxygen reserves are available for deep dives to be completely aerobic. Selective circulatory 'shut down' to non-essential structures ensures that tissues sensitive to anoxia are protected, leaving the remainder to a predominantly anaerobic metabolism. Rapid repayment of this oxygen debt is necessary on surfacing and this may explain the large tracheas and larynges found in many aquatic mammals (Figure 5.12). The long rest period at the water surface to remove lactic acid is not seen in one of the most completely adapted of all marine mammals, the northern elephant seal (*Mirounga angustirostris*). Observations show that this seal spends less than 15% of its time on the surface and dives to more than 750 m, staying down for long periods. Other pinnipeds are restricted to dives of 20 min with 3 min on the surface but the elephant seal shows no obvious relationship between depth of dive and recovery period. Most dives appear to be within the limits of individual oxygen reserves and, although no explanation is available of how this unique behaviour is possible, it may be that energy is conserved by the animal sleeping at the bottom of its dive. Here food is plentiful and the animal is safe from its prime predator, the great white shark (Cherfas, 1987).

Diving women of Korea and Japan
Humans compare unfavourably with other swimming and diving mammals in speed, manoeuvrability whilst swimming and depth of diving. Shellfish (for pearls), abalone, lobsters and seaweed around the southern shores of Japan and Korea have been collected for more than 1800 years by human divers working at depths of 10 to 25 m. These divers are mostly women, they are called Ama in both countries and can breath hold for up to 3 min during their dives. Prior to diving, the Ama hyperventilates for 10 s, takes a deep breath which removes carbon dioxide and ensures that the lungs are full of relatively rich oxygen. Compression at depth reduces lung volume by as much as half its original volume allowing oxygen transference to continue at a high ratio (Ki Hong & Rahn, 1974). Divers wear goggles or a face mask to enhance visual acuity, although this may cause hazards at depth from pressure effects on the conjunctiva. Training is essential, for although experienced women can dive repeatedly with only short periods of rest, they will suffer from periods of short anoxia. Increased vital capacity appears to be the only morphological change detected in experienced Amas and is probably the result of continued deep inspirations, for despite a high food intake these women are said to be unusually lean. As with other mammals submerged in water, reduction of heart rate to around 60 beats per minute is common, recovering to normal after surfacing.

Mammals that swim and dive appear to have common physiological adaptations that have evolved to deal with the problems that are encountered by aquatic mammals who dive to great depths or for a long period of time. The larynx and trachea appear to be larger than in terrestrial mammals of similar body weight and this may be associated with the respiratory changes that are associated with deep diving and ascent.

Locomotion at high altitude

Although the percentage composition of the atmosphere remains unchanged at high altitude, defined by Heath & Williams (1979) as above 3000 m, reduction in barometric pressure can have serious physiological effects. In spite of this, many animals, including humans, live successfully at altitudes of more than 5000 m. Notwithstanding the cold, the air temperature decreasing by one degree Celsius with each 640 feet of altitude, the prime problem is fall in ambient oxygen pressure. Hyperventilation can maintain an adequate oxygen tension within the alveolae although this results in alkalosis from loss of carbon dioxide, compensated for by renal excretion of bicarbonate.

An adequate supply of oxygen is the product of cardiac output, haemoglobin concentration and oxygen saturation of arterial blood. Full adaptation results in an increase in circulating red blood corpuscles and haemoglobin concentration. People living in the Andes have exceptionally large chests, hearts and lung volumes, the latter enhancing the amount of air taken on inspiration. Both haemoglobin and red cell concentration are increased, and the lung capillaries dilated, to increase pulmonary circulation. Metabolism and the heart rate are reduced, but, as with the Sherpa's of the Himalayas, in the acclimatized this does not effect the capacity for hard work (Hock, 1970). Other more subtle biochemical and physiological changes have been detected but in general, adjustments in permanent high altitude dwellers are similar to those acquired by 'newcomers' after about a year of residence. Acclimatization is usually maintained, although the danger of pulmonary hypertension and congestive cardiac failure remains. Carotid bodies and other oxygen sensitive tissues respond to chronic hypoxia by increasing in size and weight in both humans and other mammals living at high altitude. The incidence of tumours of these tissues is said to be at least 10 times greater in populations living in the Andes than at sea level (Heath & Williams, 1979). Chronic hypoxia also results in pulmonary arterial hypertension with an increase in thickness of the muscular wall of the pulmonary trunk. Even after long periods of acclimatization newcomers to high altitude remain more susceptible to diseases such as mountain sickness, pulmonary oedema and hypertension, infection, cold and the production of abnormal haemoglobins. The Quechua Indians who have lived in the high Andes for generations appear largely immune to such conditions, probably reflecting the elimination of susceptible genes within static populations.

Birds are more tolerant than mammals of high altitude existence although species such as the yak (*Bos mutus*) live at more than 6000 m. Cattle do not acclimatize successfully at high altitude probably because of their naturally muscular pulmonary vasculature rendering them susceptible to pulmonary hypertension. Camelids such as the guanaco (*Lama guanico*), llama (*L. glama*), alpaca (*L. pacos*) and vicuna (*Vicugna vivugna*) together with the yak, are indigenous to high altitude. All have thin-walled pulmonary arteries and haemoglobin that has a higher affinity for oxygen than that of other mammals. Concentration of haemoglobin or red cell mass is not greatly increased thus limiting the risk of pulmonary hypertension or cardiac failure. Ellipsoid shape of the red cells enhances surface area for oxygen diffusion and there is a high concentration of myoglobin in these high altitude mammals. Although the yak, living at more than 6000 m and the heaviest of these high altitude artiodactyls is slow – both vicuna and llama can run for long distances at more than 45 km per hour. Similar acclimatization is found in the deer mouse (*Peromyscus maniculatus*) which is found throughout North America as far as the Yukon. Hock (1970) trapped colonies of animals at seven locations between sea level and 4500 m, finding that with increasing altitude there was a corresponding increase in heart size. Following transference experiments between colonies he concluded that the adaptive mechanisms found in the native high-altitude mice were probably related to environmental needs rather than inherited. Native populations living in high altitudes are not genetically pure and show no morphological features, such as an unduly large laryngotracheal complex, that might be of benefit in acclimatization to environmental hypoxia. High altitude mammals, although possessing blood that appears to have a high affinity for oxygen, show no gross morphological variations from low altitude mammals of similar weight and locomotor performance (Figures 5.13 and 5.14). Successful habitation in these areas of hypoxic stress appears to be secondary to physiological adaptations shared by all species.

5.3   Vocalization

Every mammal possesses a larynx with potential for making sounds. These serve to attract, inform or warn, and are the product of natural selection (Bright,

1984). The commonest use of laryngeal sound is for communication which may be defined as 'the transmission of information that influences the listener's behaviour'. Sound, however, is not the only communication system available to mammals, as taste, smell and touch can all transmit specific useful information. Sound, however, conveys the most information per unit time, without the need

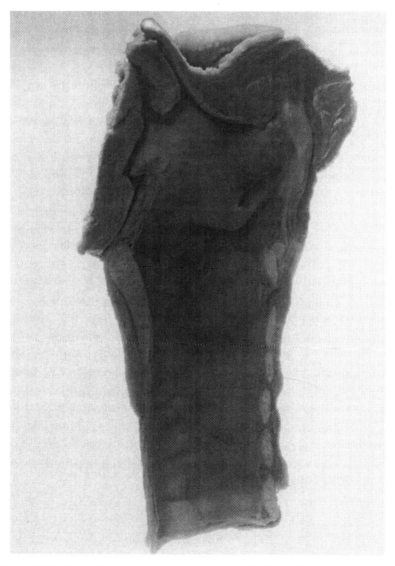

Figure 5.13. Sagittal section through the larynx of a llama (*Lama glama*) showing the short vocal cords but relatively large trachea.

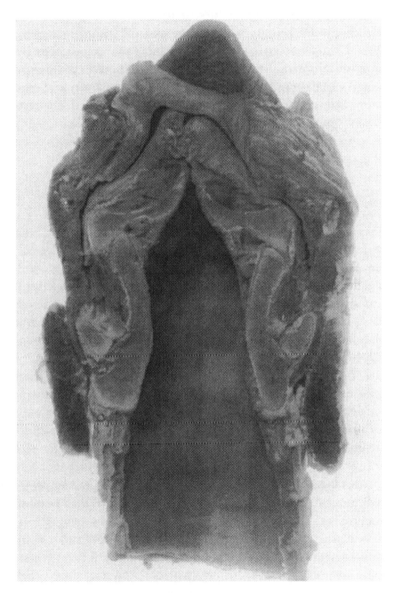

Figure 5.14. Coronal section through the larynx of llama (*Lama glama*) showing the large arytenoids and bulky aryepiglottic folds typical of a herbivore.

for direct observation as in visual signalling or attraction of predators, as in scenting. The simplest of laryngeal sounds are continuous pure tones, but to avoid interference and confusion biological sounds tend to be more complex, with overtones or harmonics (multiples of the fundamental note) spread across a specific sound spectrum. This enables each individual to produce a range of distinctive identifiable sounds, which in the case of bats may range as high as 200,000 Hz. Changes of pitch, variations in volume and temporal patterning can all convey more information than simple pure tones, this is best demonstrated in sophisticated whale songs and human speech. The complexity and significance of the latter will be discussed later in some detail; most mammals however utilize barks, grunts, honks, etc., which although relatively simple in content, contain valuable interspecific information. Indeed it might well be thought that the only mammal that uses sound unproductively is the human.

'Man', the articulate mammal (Aitchison, 1978)

In spite of general ability of mammals to produce complex sounds, humans are the only mammals to speak and understand language. Every normal human being can talk, although whether language is partially innate or wholly learnt remains controversial. There is much more to language than joining together words, and speech requires knowledge of a highly complex set of rules which are probably so fundamental, yet subtle, that children cannot acquire them by simply listening to their parents. Many linguists believe that humans are genetically predisposed to learn these grammatical rules, with crucial elements of language being held within our genes (Shanks, 1993). This is an extension of the long debate between empiricists, who argue that all knowledge is acquired, and rationalists who believe that some is inherited. Jakobson (1940) says that the sounds of all languages are composed of individual 'units', the biological bases of which are innately possessed by every human. Individual languages use subsets of these units, enabling normal children to learn any language. In spite of lengthy, wide ranging research, sometimes of debatable value, over five decades, this theory remains popular and many linguists believe that human speech involves a number of inherited neural mechanisms (Lieberman, 1991).

Many animals appear to 'talk'; parrots and mynah birds imitate humans in a variety of languages without understanding the meaning. This may also be true of attempts to teach language to chimpanzees who are not physiologically capable of uttering human sounds, despite possessing a larynx remarkably similar to humans. Most mammals, however, use sound for communication, although doubt remains as to whether human language evolved from a more primitive form of communication. Cries relating to pain or fear are common to most terrestrial mammals and exist concurrently with language in man. Indeed, vocal-auditory signals are used by many mammals, although their significance in 'communication' requires confirmation by behavioural studies.

Acquisition of speech

Evolutionary developments within the human larynx and neurological pathways allow humans to transmit and then decode, phonetic segments at a rate of between 15 and 25 per second. This is crucial to rapid vocal communication, allowing rapid transmission of complex thoughts whilst freeing the hands for other purposes. Most mammals have neurological systems that respond to their individual vocalizations. These, however, are limited in content in contrast to humans who possess the ability to produce and understand an indefinite variety of sounds. Although both non-human mammals and humans communicate by making sounds that serve common needs, there are important differences both quantitatively and qualitatively.

Language is passed from generation to generation, the child brought up in social or acoustic isolation will fail to speak. Irrespective of geographical siting, most children learn to speak by 18 to 24 months of age. This is long before there is a need for communication within this limited linguistic environment. The stimulus appears to be brain growth, although Lieberman *et al.* (1971) describing three stages in the acquisition of speech, related them to limitations in growth of the neonatal vocal apparatus. Cries recorded from 20 normal newborn human infants (some spontaneous and others 'elicited by pinches'), showed a fundamental frequency of 400 Hz; they calculated that the supralaryngeal vocal tract configuration was uniformly 7.5 cm long. Since the larynx is positioned high in the neck at this age, the vocal tract would be about half the length found in the adult. This is similar to that found in many non-human primates but they also lack a pharynx that can vary in cross-sectional area. Combined with a totally oral tongue, sound spectrograms show the formats of a supralaryngeal vocal tract shape consistent with a uniform tube open at each end. Until the larynx descends and the tongue becomes partially pharyngeal, language will therefore not mature. Most children appear to pass through similar stages of language acquisition, although differing in the age at which these levels are reached. Babbling is replaced by single and then two-word utterances, before words appear about the age of two years (Aitchison, 1978). Although much of this appears to be inborn knowledge combined with enhanced cerebral capacity, the linguistic environment plays an essential role in the acquisition of individual languages. Early in life every normal child possesses the capacity to learn any language; unfortunately this ability does not last for ever!

Speech involves continual change in the shape and length of the vocal tract in order to generate varied format frequency patterns. Comprehension requires the assignment of patterns of format frequencies, discrete phonetic categories, together with other clues to constitute words (Lieberman, 1991). Only the human vocal tract and neurological control has the ability to produce the full range of sounds necessary for speech. Whether this was acquired concurrently with an upright posture and increased cerebral capacity, or itself stimulated a bipedal posture will be discussed in the final chapter.

The physiology of voice production

Although, phylogentically, phonation is the least important function of the larynx whose prime role is protection of the lungs and control of pulmonary respiration, the unique position humans occupy amongst primates is due to the acquisition of speech. This is achieved at the cost of laryngeal protection, for the larynx has to lie low in the neck, in order to achieve both resonance and variety. Although research into voice production has previously concentrated on the animal larynx, increased interest in speech physiology and pathology has led to greater understanding of the complexities of human speech and the mechanics of voice production.

In 1741, Ferrein carried out acoustic experiments on the excised larynges of dogs by approximating the vocal cords mechanically and blowing air upwards from the subglottis. By 1837, Johannes Muller had mastered techniques for artificially producing sound from the larynges of human cadavers, noting the effect of variations in subglottic pressure on both the intensity and fundamental frequency of phonation. From this research came the classic aerodynamic and myoelastic theories of phonation.

During phonation the air stream, passing through the glottis, pulsates with a frequency which corresponds with the vibratory frequency of the vocal cords. It is this pulsating air stream which is the primary source of laryngeal sound, eventually to be modified within the supralaryngeal vocal tract. Resonation in pharynx, oral and nasal cavities, together with movements of tongue, lips and soft palate generate the changing format frequency pattern recognized as human speech (Sonesson, 1960).

The mechanism of vocal cord oscillation is influenced by energy transfer between fluid flow and vocal cord tissue vibration. Vocal cords vibrate in specific modes excited by the airstream, although self-oscillation can occur once threshold lung pressure has been reached (Titze, 1993). This is determined by the mechanical and geometrical properties of the vocal cords and is affected by tissue viscosity. Highly viscous tissues require greater lung pressure for self-oscillation, providing an explanation for adequate lubrication of the vocal cords by ventricular mucous. The viscoelastic properties of both vocal cords and ventricular folds in the excised canine larynx have been assessed by Haji *et al.* 1992). Analysing the stress-strain relationship under varying experimental conditions they concluded that cordal 'stiffness' increased with contraction of the cricothyroid muscle. This resulted in a rise in pitch. Contraction of the thyroarytenoid muscle reduced 'stiffness', increasing its viscosity. They measured the effects of these two muscles separately, although under physiological conditions they contract simultaneously in various combinations. Dry vocal cord mucosa was associated with increased stiffness, as did submucosal injection of saline to mimic cordal oedema.

The investigations of vocal cord morphology and tissue elasticity have led to the conclusion that phonation can be considered as a layered configuration of viscoelastic tissues influenced by both extrinsic and intrinsic laryngeal muscles. Vocal flexibility is provided by transglottic pressure, variation in glottic shape as well as changes in the mechanical properties of viscoelastic tissues produced by the laryngeal musculature (Titze & Talkin, 1979). The inaccessibility and complexity of this system, however, restricts direct measurement of many of these variables; particularly glottic volume velocity which is important in speech (Koyama *et al.*, 1969).

Electromyography has been used to obtain electrophysiological data of both normal and functional disorders within the larynx (Thumfart, 1988). This confirms that the fundamental frequency although scaled according to laryngeal size and length, is controlled by the cricothyroid muscle. This is of importance in vocal training where singers are attempting to produce varying fundamental frequencies within their normal range. As higher fundamental frequencies are attempted, a greater range of lung pressures are needed to obtain a similar level of sound pressure. This is best assessed by Voice Range Profiles, in which sound pressure levels in dB are plotted against fundamental frequency (Titze, 1993).

Effective production of a normal voice requires the co-ordination of a variety of differing factors; subglottic air pressure, glottic air flow and resistance, muscular contraction, vocal cord stiffness and acoustic coupling of the supralaryngeal and subglottic spaces. Final articulation results from modulation of sound by tongue, lips, palate and teeth, each playing a specific role in differing languages (Wood, 1986). Increasing interest in the human voice has generated more research and a need for standardization of these parameters (Gould & Korouin, 1994). From measurements of vital capacity; tidal volume, expiratory flow volume and inspiratory capacity can be calculated and compared with predicted values. Laryngeal function can then be evaluated by visual and acoustic analysis and aerodynamic studies. Stroboscopy with video or photographic storage is now a standard technique, whilst sound spectrography allows analysis of frequency, amplitude and phase relations or other parameters of voice. Airflow measurements by pneumotachography indicate glottic closure efficiency, whilst electroglottography records cycle-to-cycle vocal cord function. Electromyography allows objective analysis of laryngeal muscle activity, whilst studies of the spatial, aerodynamic and acoustic characteristics of the supralaryngeal tract are possible using radiological techniques (Hirano, 1981), Titze (1994) has described the need for adopting standardization of acoustic voice analysis as being educational and necessary for simplification of a potentially confusing situation. Absence of specificity and universal acceptance of test material restricts the value of subsequent acoustic analysis. Without it, technical criteria cannot be established for estimating fundamental frequency, intensity, spectral analysis and other parameters. Even then it is doubtful if an individual's potential to speak or sing with exceptional ability can be anticipated by purely physiological investigations.

Voice in the elderly

There is evidence that the voice changes with increasing age, this is more noticeable in professional singers and other professional voice users. Honjo & Isshiki (1980) studied the laryngoscopic and voice changes using voice recordings in 20 Japanese men and 20 women with a mean age of 75 years. They found vocal cord atrophy and oedema in most of this group, with increased fundamental frequency in males but reduction in females. Vocal pitch, as expressed by the fundamental frequency, was 162 Hz for speech in aged (mean age 75 years) men (young men 120–130 Hz) and in women 165 Hz (young women 260 Hz).

Both Mysak (1959) and Hollien & Shipp (1972) reported similar findings, concluding that changes in fundamental frequency with age were related to fatty degeneration of the vocal cords. Oedema was more common in the cords of women and this increase in bulk may lower the vocal pitch. These investigations failed to consider the possibility of other factors, such as muscle weakness or reduction in respiratory function playing some role in the generally accepted change in voice that may come with increasing age.

Professional voice

Although all normal humans can speak few possess the innate ability to excel in speech or singing that is accepted by the society in which they live. Such attributes are not necessarily universally attractive, since both language and musical tastes vary considerably among humans. The structure of the larynx in these individuals in no way differs from that found in other healthy normal humans, many of whom may be lacking motivation or dedication. The qualities of structure and function necessary for 'fine' singing will not differ greatly from those needed for professional speaking, except that actors are allowed discretion in the manner of pitch and pace, but less with regard to articulation (Punt, 1979). The greater the vital capacity the better the singer can sustain a long phrase or limit the number of breaths taken. In unpublished research on a group of young trainee singers followed up over a period of four years, the only physiological change detected by Harrison was an increase in vital capacity. Good lung function is of course essential, requiring normal lungs and good chest and diaphragmatic muscles. Obviously, the larynx must be absolutely normal with vocal cords free from any pathology, particularly the cordal edge. Laryngeal size will affect the fundamental note, this being particularly relevant in a voice of low tessitura (bass or contralto). One of the most important factors, however, is intrinsic lubrication by ventricular mucous. A dry environment will have a dramatic (even disastrous) effect on note production (Pressman & Keleman, 1955).

The role of resonance, although appreciated by both singer and coach, remains uncertain. It appears likely that the nasal passages and paranasal sinuses play only a minimal role, except when a source of infection. The pharynx and oral cavity are considered to be of paramount importance, although the degree of flexibility and size of these cavities cannot be measured objectively or greatly influenced by training programmes. Those individuals who have been provided by nature with pharynges and oral cavities whose dimensions, openings and tissue properties produce harmonic partials of richness of timbre or clarity of words are unusually fortunate. 'Great' singers or speakers must therefore be considered as unusual variations of structures primarily designed for lung protection and respiration. It is this natural ability that can be improved by skilled training; although intrinsic deficiencies in vocal intensity, range or pitch discrimination cannot be as readily improved as breath control.

Many mammals possess capacious pharynges and oral cavities. Lions and seals have capacious pharynges and oral cavities, producing loud sounds of wide frequency range. Although an intranarial epiglottis and wholly oral tongue restricts the range of sounds produced by all mammals except humans, sound spectrometry reveals that many utilize complex patterns of vocalization for communication, territorial identification, mate seeking or even in some chiroptera and cetacea, the catching of food. Speech as recognized within the human environment appears to be unnecessary for their natural survival; its role in the evolution of man will be discussed in the final chapter.

Comparative vocalization

Although all mammals have the ability to make sound, many appear to be aphonic except under conditions of extreme stress. This is particularly true of rodents and lagomorphs, despite many of these animals having long sharp-edged vocal cords. If the concept that communication by sound has evolved to help the sender to influence the behaviour of the recipient is accepted, then, vocalization has to be considered in relation to patterns of behaviour. Human language may be considered as the ultimate 'communication device', but it should be appreciated that many mammals use sound to identify specific dangers or for explicit purposes. Analysis of sound recordings made under natural conditions has provided data as to the frequency composition of these emissions, allowing correlation with behaviour perspectives.

Many mice, rats, gerbils and hamsters are known to produce ultrasonic as well as sonic sounds (Sales, 1970). Detailed studies of the significance of ultrasonic vocalization in seven species of *Rattus* native to Australia, together with the black rat (*Rattus rattus*) and brown rat (*R. norvegicus*) have been reported by Watts (1980). Both audible and ultrasonic sounds were recorded and the ontogeny of the different calls investigated. This provided an overview of the range of vocalizations produced by different species of rats. Four distinct classes of sound were identified: squeal, ultrasonic pipping, ultrasonic whistle and coughing. Each of the first three vocalizations had two distinct forms, giving a total of seven discrete calls. Despite the considerable phylogenetic and ecological diversity between these species, the overall findings with respect to vocalization were found to be remarkably similar. All nine species had the same basic repertoire of calls, using them in similar situations, suggesting that these sounds play a fundamental role in the behavioural pattern of *Rattus*. A similar evolutionary conservatism of vocal repertoires has been found in Australian hopping mice (*Notomys* sp.) by Watts (1975) and in some ungulates by Kiley (1972).

Ultrasonic sounds of 50,000 Hz emitted by adults were particularly noticed in aggressive situations, falling to 22,000 Hz during submissive postures. Babies produced ultrasonic calls to attract their mothers, although these appeared to be inadequate for accurate identification without olfactory assistance. Other rodents such as the woodmouse (*Apodemus sylvaticus*) and some shrews also emit ultrasonic sounds, but use them for varying purposes, such as echolocation. Examination of the larynges of these species shows no gross or histological features which could be correlated with an ability to produce ultrasonic sound. It is possible that many mammals have this ability but find it unnecessary.

The need for the young animal to recognize and be recognized by its mother, often without attracting the attention of predators, is fundamental to survival for most species. Ultrasonic emissions can achieve this, although piglets and lambs recognize the voice of their mothers, who wait for their young to come to them. Sonographic analysis of vocalizations from ewes, however, show individual variations within this species, with some having homogeneous voices. Such variations in behaviour patterns indicate that, in non-human mammals, vocalization is complex and diverse.

South American tapir (Tapirus terrestris) *and giant panda* (Ailuropoda melanoleuca)

There is an obvious need for an effective means of communication among mammals living in herds or groups but it is less clear among those that lead a relatively solitary existence.

Tapirs are solitary, not usually found in groups except in captivity. Hunsaker & Hahn (1965) studied the vocalization of eight tapirs living in a large compound within San Diego Zoo. Four sounds were detected in their social interactions of which two were laryngeal in origin; the remainder were a nasal snort and a lingual-palatal click. The most distinct sound was a squealing call uttered during pain, fear, appeasement and exploration. The squeals ranging in intensity from 10 to 135 dB and although related in part to age, were most commonly influenced by emotional stimuli such as fear or anger.

Tapirs live in dense jungle precluding easy visual location and identification of predators and other tapirs. Vocalization is of supreme importance for communication with partners during the breeding season, for olfaction is only useful in situations of close proximity. The limited range of laryngeal sounds is adequate for the restricted needs of this species; low intensity for local communication whilst the high intensity squeal of fear or pain serves as a warning to animals further away. The larynx is similar to that of other herbivores (Figure 5.15). The giant panda lives a largely solitary existence except when courting or during accidental social interaction. Peters (1985) studied tape recordings from seven adult animals (six females and one juvenile male), relating specific sounds to behaviour patterns. These relationships have also been studied in pandas in their natural environment by Schaller *et al.* (1985). Eleven discernible sounds were identified of which eight emanated from the larynx; huffing, snorting and chomping were thought to be non-vocal. The honk, a short call emitted singly or repeatedly, had its main frequency between 0.2 and 2.7 kHz with one or more harmonics. This sound appears to indicate mild distress.

Bleating is similar to that of the goat rather than sheep, and is described as a 'tonal call of medium intensity with rapid frequency and amplitude modulation' (Schaller *et al.*, 1985). Main frequency range is up to 3 kHz with several harmonics, the bleat being best heard during the mating season. Moaning is a highly variable vocalization, with a prime frequency below 1 kHz; this was the most common vocalization. The panda's bark is similar to that of the dog, with a prime frequency below 1.5 kHz. In the wild the panda barks when courting, startled or casually meeting other animals.

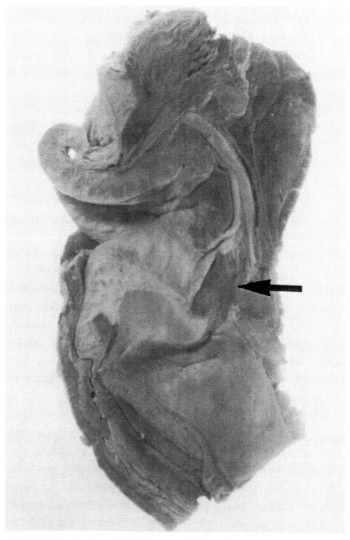

Figure 5.15. Sagittal view of the larynx of a Brazilian tapir (*Tapirus terrestris*) to show the bulky but short vocal cord (arrowed).

Chirping can be surprisingly high-pitched with frequencies reaching 8 kHz or as low as 1 kHz. Its use appears to be similar to the bark and observed animals switched from one to the other as emotional levels changed. Squealing and roaring are high intensity calls of variable tonality and duration, usually heard during serious fighting. The volume of sound produced in roaring is surprising and indicates an aggressive, dominant animal.

In order of increasing intensity the moan, bark and chirp appear to represent a graded system of emotional change. Pandas use these sounds randomly, one interchanging with another and their function varying with circumstances. Despite their relatively lonely existence and 'carnivorous'-type larynx the giant panda has a varied repertoire of vocal sounds. Schaller *et al.* (1985) compared these sounds with those made by bears, raccoons and the red panda. They emphasized that the sound spectrographic analysis has to be viewed in a functional context, for sounds of similar structure may not have the same functional significance. Whereas the giant panda's honk structurally resembles the grunt of bears and raccoons, in the former it denotes stress but in the latter, social contact. They concluded that the vocal repertoire of giant panda, black bear, raccoon and ringtail possum were remarkably similar; the red panda appearing to have fewer vocal signals. Most agonistic sounds – huff, snort, growl, roar and squeal – appear to be widespread among many mammals, and is unrelated to body size, habitat or laryngeal structure.

Roaring and non-roaring cats

In mammals the hyoid bone is suspended from the skull base by a variety of small bones or ligaments. Owen (1834) dissecting an adult lion found an incompletely ossified hyoid with an elastic ligament six inches (*c.* 15 cm) long replacing the epihyal bone. This he believed allowed the larynx to be moved away from the palate thereby lengthening the pharynx. Confirmation of this finding in other large cats (*Panthera* spp.) was reported by Pocock (1916) and this was accepted as the explanation for 'roaring' in these animals.

Hast (1989) dissecting the larynges of 26 members of the family Felidae showed convincingly that in lion, tiger, leopard and jaguar there was a pad of fibro-elastic tissue in the rostral part of their large vocal cords. This was not found in the Felidae who have divided thyroarytenoid muscles, ventricles and sharp-edged vocal cords. Sections of this pad of tissue showed collagenous and elastic fibres, that were denser near to the epithelial mucosal lining. He concluded that because of their large mass, the *Panthera* spp. vocal folds generate sounds of low natural frequency with high acoustic energy (Figures 5.16 and 5.17). This genus have large cricothyroid and vocalis muscles

(Appendix 2) and with their capacious pharynges and mouths appear ideally suited for high volume vocalization. The snow leopard (*Panthera uncia*) is unique amongst the *Panthera* spp. in that although its larynx is similar to the jaguar, it has no fibroelastic pad and does not roar (Figure 5.18). Hast (1989) feels this is enough to justify the classification of this species as a separate genus *Uncia uncia*.

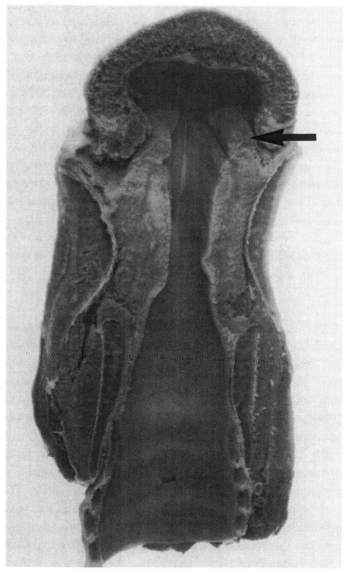

Figure 5.16. Coronal view of the larynx of lion with the thick fibroelastic supra arytenoid mass (arrowed) thought to be responsible for 'roaring.'

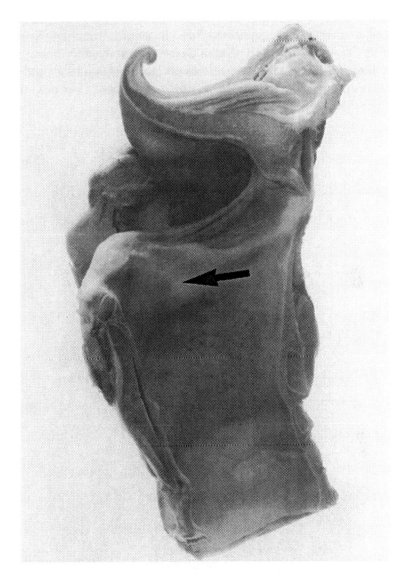

Figure 5.17. Sagittal view of the larynx of a Siberian tiger (*Panthera tigris*) showing similar fibroelastic tissue to that arrowed in Figure 5.16.

Both the puma (*Felis concolor*) and cheetah are of similar size to the leopard, but have short sharp-edged vocal cords. They purr, growl or scream but do not roar.

At the time of the breeding season (the rut) dominant red deer (*Cervus elaphus*) stags bellow and roar, despite normally being animals with minimal vocalization. During the three weeks that are spent bellowing and defending their harem from interlopers, the stags stop feeding although consuming considerable energy. There is evidence that those stags which prove to be the best fighters are those which have the loudest and most repetitive roars; possibly because both are related to physical fitness. Stallions with the loudest and longest-lasting squeal are likely to be the best fighters, because this sign

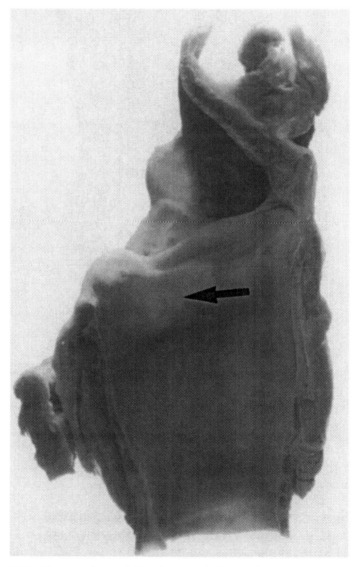

Figure 5.18. The snow leopard (*Panthera uncia*) does not have this fibroelastic pad and does not roar. The area shown in Figure 5.17 is arrowed in this sagittal view.

indicates a large lung capacity. Squeals of dominant stallions have been timed as lasting 20% longer than in subordinate animals and are at a higher frequency.

In the case of the red fox (*Vulpes vulpes*) the screams emitted by both dog and vixen are more complex in their significance. Primarily designed for long distance communication (they can peak at more than 160 Hz) they can also be specifically directed to individual receivers. Less audible sounds for short distance purposes such as alarm calls, indicate that like so many other mammals the full pattern of vocalization is complex and variable.

The volume and frequency variation of mammalian vocalization depend upon laryngeal and lung capacity and the length and thickness of the vocal cords. However, the pattern of sounds produced is clearly allied to functional require-ments and as such could be considered as equivalent to human language.

Primate vocalization

Although human language is unique in allowing the transmission of abstract thoughts such as events that have yet to happen, it would be erroneous to assume that other primates utilize a primitive form of vocalization. Research into the calls of monkeys confirms that those living in dense forest habitats require complex patterns of vocalization for communication. Pola & Snowdon (1975) studied the pygmy marmoset (*Cebuella pygmaea*), an arboreal species, living in the South American tropics. Ten calls were recognized, each relating to identifiable situations, but as with other correlation studies there were fine distinctions between separate vocalizations. Identification of specific behaviour patterns relating to identifiable sounds is particularly difficult in dense forest, although these ten sounds are similar to those found in studies carried out on other species of marmosets.

Howler monkeys (*Alouatta* spp.) also live in the rainforest but have a call as loud as a roaring leopard, produced within a larynx which has an expanded thyroid cartilage and long vocal cords. Early morning vocalization serves to identify the feeding sites of individual groups so avoiding exhausting confronta-tion with other bands (Mitchell, 1986). Loud calls are also given by dominant male mangabeys (*Cercocebus atys*) although in this instance both frequency and timing vary with individual males.

Possibly the most elaborate and loud sounds are produced by both sexes of gibbons and have been recognized for more than 2000 years, when they appeared in Chinese poems, legends and myths (van Gulik, 1967). Haimoff (1984) has studied the vocal repertoire produced by various species of gibbon (*Hylobates* spp.) finding a narrow frequency bandwidth and relatively low frequency range; all ideal for long distance propagation of sound (up to 3 km in dense rain forest). Some sounds appear to indicate age, sex and identification of the caller and gibbons of different species have distinctive, recognizable calls.

The vervet monkey (*Cercopithecus aethiops*) is however, of the greatest interest since there is evidence that it has distinct alarm calls for different types of dangers. Adult animals give different alarm calls depending on which of three predators are sighted: python, martial eagle or leopard. Seyfarth *et al.* (1980) showed that when tape recordings of each of these calls is played, adults respond in differing and appropriate ways. These alarm calls are not entirely predator specific in that leopards, lions, hyaenas and cheetahs will all trigger the 'leopard response'. It also appears that these responses have to be learnt, since infants give leopard alarm calls to warthogs or hawk calls to pigeons.

Vervet monkeys have a complex social system with calls serving significant and varied functions. Sonographic analysis of their grunts show many differences each of which serves a specific purpose. Together with their specialized alarm calls, this species appears to have developed a sophisticated system of communication not entirely dissimilar from language.

Great apes The close genetic relationship between humans, gorilla, orang-utan and chimpanzee suggests that similarities may exist within their patterns of vocalization. Macroscopically, ape larynges differ from those of humans in size, the presence of ventricular air sacs and an intranarial position. The latter is associated with a restricted supralaryngeal airway and curtailed articulation.

As with other mammals, patterns and significance of vocalization appear to be associated with habitat and life-style. Both male and female orang-utans live solitary existences meeting only by accident or when the female is in oestrous. Male animals broadcast their presence by a low, sliding moan which builds to a powerful bellow, possibly amplified by the ventricular air sac. This sound can be heard for several kilometres and serves as a territorial warning to other males as well as attracting passing females.

Fossy (1972) collected data from seven gorilla groups totalling 106 animals, living in their natural habitats. Sixteen types of vocalization, which were then grouped into 12 basic categories, were recognized. The aggressive calls included roaring, growling and panting; mild alarm calls were variations on barks whilst fear produced screams or silence; and distress produced cries or whines, although many variations of wails, sobs and screeches were heard in young animals (Figure 5.19).

Group vocalizations including grunts, hoot barks and chuckles contributed to a varied and complex series of vocalizations, which appear to be expressions of a

Figure 5.19. Sagittal view of the larynx of the gorilla showing its simularity to that of humans.

wide range of moods. Although the mature silverback was found to call most frequently, possess the largest sound repertoire and use non-vocal sounds such as chest beating; other members of the group contributed, particularly with the less aggressive sounds.

Vocalizations play an equally important part in the chimpanzee's social repertoire, although reinforced by facial expression and other non-vocal be-haviour patterns. Passingham (1982) describes 13 individual calls, although the pant-hoot, a loud call uttered most frequently by the male and carrying for several metres in dense forest, appears to be the most frequent. The basic expressions of threat, submission and play are similar to those found in other primate groups. The pant-hoot starts with a low-frequency sound that builds in intensity, with increasingly rapid breathy inhalations and exhalations. The climax is a high-pitched scream after which the calls cease. (Small, 1994). Chimpanzees live in fluid communities of females and males, with individuals spending much of their time alone although congregating for feeding and social interaction. Vocalization appears essential for social survival with young animals having to learn the group repertoire soon after birth. It is clear that the various sounds are both subtle and complex in their social implication and may thus be similar in impact to human language.

Despite the environmental restrictions inherent in the correlation between mammalian vocalization and behaviour patterns, there appears to be little doubt that most mammals produce laryngeal sound for specific purposes. Spectroscopic analysis is revealing the complex structure of these sounds, which field studies are attempting to associate with identifiable response patterns. Although influenced by resonation from supralaryngeal structures and possibly large air sacs, these sounds are produced within the larynx. Some Cetacea and Chiroptera, however, possess the ability to produce laryngeal ultrasonic sound for echolocation, navigation and food capture.

Echolocation in the Cetacea and Chiroptera

Odontocetes (toothed whales) vocalize for communication and echolocation. They hunt by sight in clear water, by detecting noises made by potential prey or by echolocation; for they can emit a wide range of sounds from their larynges or possibly nasal region. Aristotle described the moans and squeaks made by dolphins (Chapter 1) and with their large brains many odontocetes have been credited with powers of communication and thought not dissimilar to man. Sound emissions are classified into two general categories: pulsed sounds which have been described as yelps, squawks, squeaks, groans, barks and clicks. The second type are a family of frequency modulated pure tone whistles which are thought to be primarily for communication purposes (Gaskin, 1982). Sonographic analysis confirms that pulsed click-sounds are used for echolocation and navigation, although pure tone whistles have been recorded simultaneously with clicks, suggesting the possibility of a dual origin. As with other mammalian vocalization, the odontocetes, which include dolphins, porpoises and a variety of whales, radiate a spectrum of sounds that must be viewed in the context within which they are emitted. Some of the pure tone whistles may even be specific to individual animals (Caldwell, 1968).

Most toothed whales appear to use echolocation (a form of autocommunication) to search their environment for obstacles as well as food. The bottle-nose dolphin (*Tursiops truncatus*) can do this in complete darkness using clicks varying between 0.25 and 220 kHz, with a beam width narrow enough to stun or even kill prey. The mechanism with which these mammals produce sound for echolocation and communication has been controversial since Below (1551) first described the cetacean larynx. He failed to find vocal cords, an omission perpetuated by successive anatomists and zoologists. This resulted in the source of sound production being related to passage of air past structures such as the nasal sacs or plugs which are prominent features of cetacean respiratory physiology. The large fatty body, the 'melon', situated on the dolphin forehead is thought to focus this nasal sound to a point about 1 m ahead (Evans, 1973).

Reidenberg & Laitman (1988) investigating the laryngeal anatomy of 24 odontocetes representing 10 genera, found that vocal cords were present in all specimens together with false cords and mucus secreting ventricles. The orientation of the cords was vertical, parallel to the airflow and clearly a potential source of sound. However, ultrasound scans have suggested that with clicking sounds there is vibration of both nasal plug and related air sacs and it remains possible that some sounds are nasal in origin (Figure 5.20).

Both true and false cords contained large amounts of collagen and elastic tissue and within the large ventricles were a number of small air sacs filled with mucus. Although it is not possible to study the function of the vocal cords in living animals there appears to be no reason why the odontocete larynx should not produce sound in a manner similar to that found in terrestrial mammals. The nasal plugs and sacs surrounding the blow-hole may then act to modify this sound, much as the tongue does in human speech.

It is not surprising that, during more than 50 million years of evolution, some species of bats (e.g. Microchiroptera) have developed ultrasound as a sophisticated means of navigation and prey capture. The Italian naturalist Lazaro Spallanzani (1932) showed that blinded bats could capture insects, although it required sonographic analysis to confirm that this was carried out using ultrasound. Bats do emit sounds at about 12,000 Hz, audible to the human ear, although those used for echolocation lie between 20 and 210 kHz. Although the bat larynx lies high in the nasopharynx some species emit these sounds through the mouth, others the nose. The intensity can be as high as 200 dynes per cm^2 with both frequency-modulated and constant frequency pulses being used, depending on species and circumstances. Bats belonging to the subfamily

Vespertilioninae (*c.* 350 species) are frequency modulators producing sound with a wide band width, in short pulses of between 0.5 to 10 milliseconds and sweeping through at least one octave (Vaughan, 1986). This is ideal for information gathering of size, speed and direction of potential insect prey. The horseshoe bat (family Rhinolophidae) emits long-duration constant frequency pulses of a narrow band width. Although not conveying as much information

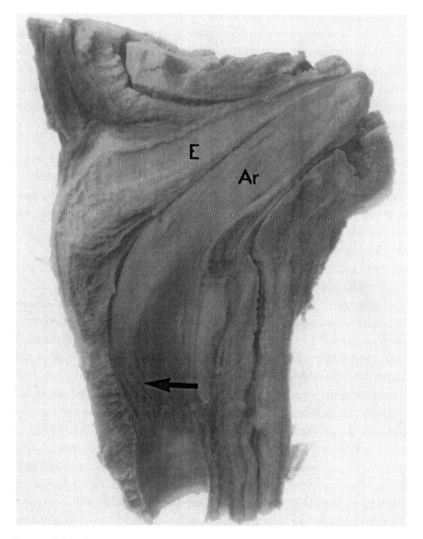

Figure 5.20. Sagittal section through the laryngeal area of a common dolphin (*Delphinus delphis*) showing the epiglottis (E), arytenoid (Ar) and vocal cord (arrowed).

about a target's characteristics, these emissions are important in detecting speed and direction of insect movement.

It now appears that bats may use echolocation systems to suit specific situations, switching from constant to frequency modulation as circumstances change. There are, however, over 950 species of bats most of which belong to the suborder Microchiroptera. Only a small number of these have been studied in detail, and the relationship of sound emission to behaviour appears to be unusually complex. Whether frequency modulated or of constant frequency, sound is produced within the larynx which in the bat is composed of ossified cartilage, large intrinsic muscles and in some species contains laryngeal or tracheal air sacs. Denny (1976) examined serial sections of the larynges from 14 families of Microchiroptera finding cricoid or upper tracheal diverticulae in rhinolophids, hipposiderae, nycteridae and one species (*Pteronotus* sp,) of phyllostomatoidae. All were constant frequency pulse nasal sound emitters with nose-leafs, except for *Pternotus* sp which uses its mouth and whose larynx appears to lie low in the nasopharynx. It is possible that these capacious air containing rigid walled cavities (Figures 5.21 and 5.22) act as resonators, but experimental evidence is lacking to confirm this hypothesis. Only a small number of species have been examined and clarification awaits further anatomical and acoustic research in other species of Microchiroptera.

Cats purring and dogs barking

A specialized form of vocalization has been recognized in the tiger for many years termed 'prusten', characterized by a short, low intensity snorting sound generated by friendly contact. A similar sound is found in the leopards, jaguar and lion according to Hemmer (1981). Peters (1984) recorded prusten in the clouded leopard (*Neofelis nebulosa*), snow leopard (*Uncia uncia*), tiger (*Panthera tigris*) and jaguar (*P. onca*) and carried out spectrographic analysis within the range of 50 to 7000 Hz. Prusten was produced during expiration and was associated with a clearly visible vibration in the region of the larynx. In several tigers, palpation of the larynx confirmed the source of vibration and concurrently fingers in the nostrils demonstrated jets of air. It was concluded that the larynx produced the low frequency component of prusten pulse sequence whilst the higher frequencies came from air expelled through the nostrils. Variations in sonographic analysis of the sounds produced by the four species of Felidae reflects structural differences in the dimensions of pharynx and larynx. A similar vocal response to pleasure or contentment is common in many cats, civets, genets, ocelots, pumas and cheetahs. Purring is a soft buzzing sound with a

fundamental frequency of between 23 and 31 Hz. Varying with respiration it is
continuous in inspiration and expiration and the intensity and duration relate to
the animals state of arousal (Sissom & Rice, 1991). Surface vibration can be
localized to the region of the larynx and EMG has confirmed that the source of
the sound is from the vocal cords (Remmers & Gautier, 1972). Variable filtering
within the supralaryngeal region can produce the rich harmonics and variations
in quality and loudness which are observed in cats of differing sizes and species.
Stimulation of regions within the brain can elicit purring, which appears to be a

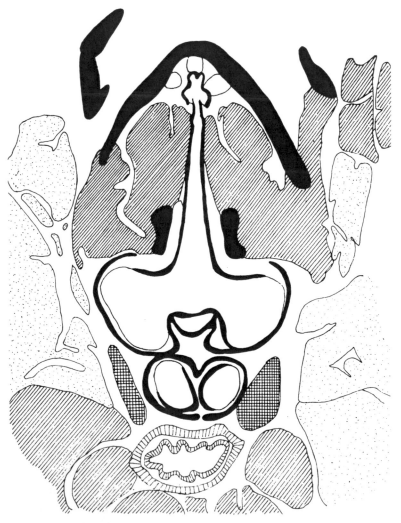

Figure 5.21. Drawing of a transverse section through the larynx of a greater
horseshoe bat (*Rhinolophus ferrumequinum*) to show the large posterior air sacs.

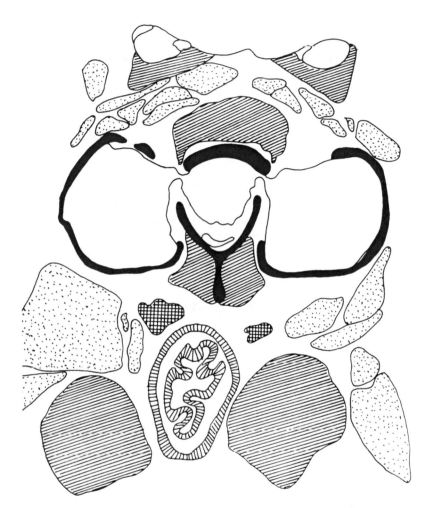

Figure 5.22. Drawing of a transverse section through the larynx of the slit-faced
bat (*Nycteris macrotis*) to show paired cricoid air sacs.

centrally driven periodic laryngeal modulation of respiratory flow. Some additional voicing is possible during expiration and this has been recorded in the
cheetah by Sissom & Rice (1991). Purring appears to be similar to prusten in
that the sound is laryngeal in origin, modified by supralaryngeal structures and
signifies contentment.

Wild dogs, including wolves, howl but seldom bark, except when young.
Analysis of sounds in a bark suggest that it lies between a puppy's distress call
and an aggressive growl. This unwarranted use of energy is thought by Messel
(1993) to represent a juvenile form of begging and therefore the result of

domestication. A sheepdog is recorded as barking non-stop for seven hours, a singular reward for being 'man's best friend!'

Every mammal possesses the ability to produce laryngeal sounds; many species show complex patterns of vocalization that can be related to identifiable systems of behaviour and communication. Although anatomical and neurological inadequacies prevent mammals from formulating the complex sounds that humans recognize and use as speech, they can serve to provide sophisticated systems of communication. Data are available about the structure of many of these sounds, and detailed correlation with individual and group responses depends on meticulous observation in the natural habitat. Environmental restrictions, such as dense rain forest, limits observation which itself is time consuming. From our present knowledge there appears to be good reason for concluding that all mammalian vocalization is significant at the species level and could reasonably be compared biologically with human language.

References

5.1 Respiration

Altman, P. L. & Dittmer, D. S. (1961). *Blood and other Body Fluids*, p. 39. Federation of American Societies for Experimental Biology.

Bruton, C. D. (1885). The valvular effect of the larynx. *Journal of Applied Physiology*, 7, 62–71.

Burke, J. D. (1966). Vertebrate blood oxygen capacity and body weight. *Nature*, 212, 46–8.

Dedo, H. H. (1970). The paralysed larynx; an electromyographic study of dogs and humans. *Laryngoscope*, 80, 1455–517.

Fink, B. R. (1978). Energy and the larynx. *Annals of Otology, Rhinology and Laryngology*, 87, 595–606.

Fink, B. R. & Demarest, R. R. (1978). *Laryngeal Biomechanics*. Harvard University Press.

Fukuyama, T., Umezaki, T. & Shin, T. (1993). Detection of laryngeal sensory evoked potentials (LSEP) in the cow. *Head and Neck Surgery*, 109, 748–52.

Gehr, P. (1981). Design of the mammalian respiratory system. V. Scaling morphometric pulmonary diffusing capacity to body mass. Wild and domestic animals. *Respiratory Physiology*, 44, 61–86.

Harrison, D. F. N. (1980). Biomechanics of the giraffe larynx and trachea. *Acta Otolaryngoloica (Stock.)*, 89, 258–64.

Hast, M. H. (1967). Mechanical properties of the vocal fold muscles. *Practical Otolaryngology*, 29, 53–8.

Jones, J. D. (1972). *Comparative Physiology of Respiration*. London, Edward Arnold.

Jurgens, K. D, Bartels, H. & Bartels, R. (1981). Blood oxygen transport and organ weights of small bats and small non-flying mammals. *Respiratory Physiology*, 45, 243–60.

Kleiber, M. (1932). Body size and metabolism. *Hildergardia*, 6, 315–53.

Konrad, H. R, Rattenborg, C. & Kain, M. L. (1984). Opening and closing mechanisms of the larynx. *Otolaryngology – Head and Neck Surgery*, 92, 402–5.

Loudon, A. (1986). Mammals produce some metabolic surprises. *New Scientist*, 26th June, p. 36.

Maina, J. N. (1987). The morphology and morphometry of the adult normal baboon lung (*Papio anubis*). *Journal of Anatomy*, 150, 229–45.

Maina, J. N. & King, J. (1982). Allometric comparisons of some morphometric parameters of avian and mammalian lungs. *Journal of Physiology (Lond.)*, 330, 10–38.

McCaffrey, T. V. & Kern, E. B. (1980a). Laryngeal regulation of airway resistance. One. Chemoreceptor reflexes. *Annals of Otology, Rhinology and Laryngology*, 89, 209–14.

McCaffrey, T. V. & Kern, E. B. (1980b). Laryngeal regulation of airway resistance. Two. Pulmonary reflexes. *Annals of Otology, Rhinology and Laryngology*, 89, 462–6.

Morrison, P. R., Pierce, M. & Ryser, F. A. (1959). Food consumption and body weight in the masked and short tailed shrew. *American Nature*, 57, 493–500.

Paintal, A. S. (1973). Vagal sensory receptors and their reflex effects. *Physiology Review*, 53, 159–227.

Patterson, J. L., Iet al (1965). Cardiorespiratory dynamics in the ox and giraffe, with comparative observations on man and other mammals. *Annals of the New York Academy of Sciences*, 127, 393–413.

Petcu, L. G. & Sasaki, C. T. (1991). Laryngeal anatomy and physiology. *Clinics in Chest Medicine*, 12, 415–23.

Rubner, M. (1883). Uber den Einflus der Korpergrosse auf Stoff-und-Kraftwechsel. *Journal of Biology*, 19, 535–62.

Sasaki, C. T. & Buckwalter, J. (1981). Laryngeal function. *American Journal of Otolaryngology*, 5, 281–91.

Sasaki, C. T., Fukuda, H. & Kirchner, J. A. (1973). Laryngeal abductor activity in response to varying ventilatory resistance. *Transactions of Academy of Opthalomology and Otolaryngology*, 77, 403–6.

Schmidt-Nielsen, K. (1984). *Scaling – Why is Animal Size so Important?* Cambridge University Press.

Stahl, W. R. (1965). Organ weights in primates and other mammals. *Science*, 150, 1039–42.

Stransky, A., Szereda-Przestaszewska, M. & Widdicombe, J. G. (1973). The effects of lung reflexes on laryngeal resistance and motorneurone discharge. *Journal Physiology (London)*, 231, 417–38.

Tenny, J. M. & Bartlett, D. (1967). Comparative quatitative morphology of the mammalian lung and trachea. *Respiratory Physiology*, 45, 243–60.

Tenny, J. M. & Remmers, J. E. (1963). Comparative quatitative morphology of the mammalian lung: diffusng area. *Nature*, 197, 54–6.

Weibel, E. R. (1979). Oxygen demand and the size of the respiratory structures in mammals. In *Evolution of Respiratory Processes*, ed. C. Lenfant, & I. Wood, p. 320. New York, Dekker.

Woodson, G. E., Ambrogio, F. S. & Mathew, O. (1989). A contraction on laryngeal resistance and glottic area. *Annals of Otology, Rhinolgy and Laryngology*, 98, 119–23.

5.2 Locomotion

Altmann, S. A. (1987). The impact of locomotor energetics on mammalian foraging. *Journal of Zoology (Lond.)*, 211, 215–25.

Bennett, M. B. (1987). Fast locomotion of some Kangaroos. *Journal of Zoology (Lond.)*, 212. 457–64.

Bonner, J. (1994). They overheat horses, don't they? *New Scientist*, 26th March, 21–4.
Bramble, D. M. & Carrier, D. R. (1983). Running and breathing in mammals. *Science*, 219, 251–6.
Brazier Howell, A. (1944). *Speed in Animals*. University of Chicago Press.
Calder, W. A. (1984). *Size, Function and Life History*. Harvard University Press.
Cherfas, J. (1987). Elephant seals reach depths other seals can't. *New Scientist*, 3rd September, 36.
Cook, W. R. (1976). Idiopathic laryngeal paralysis in the horse. A clinical and patholological study with particular reference to diagnosis, aetiology and treatment. PhD thesis, University of Cambridge.
Cook, W. R. (1989). *Specifications for Speed in the Racehorse*. USA, Russell Meerdink Co.
Corbett, G. B. & Hill, J. E. (1991). *A World List of Mammalian Species*. 3rd edn. Oxford University Press.
Day, H. (1981). *Vertebrate Locomotion*. London, Academic Press.
Dawson, T. J. (1983). *Monotromes and Marsupials: the Other Mammals*. London, Edward Arnold.
Dawson, T. J. & Taylor, C. R. (1973). Energetic cost of locomotion in kangaroos. *Nature*, 246, 313–14.
Denny, S. P. (1976). Comparative anatomy of the larynx. In *Scientific Basis of Otolaryngology*, ed. R. Hinchcliffe & D. F. N. Harrison, p. 536. London, Heinemann.
Dunbar, R. (1985). The race to breed faster horses. *New Scientist*, 6th June, 44–7.
Elsner, R. & Gooden, B. (1983). *Diving and Asphyxia*. Cambridge University Press.
FitzGibbon, C. D. & Fanshawe, J. H. (1989). The condition and age of Thomson's gazelles killed by cheetahs and wild dogs. *Journal of Zoology (Lond.)*, 218, 99–107.
Gans, C. (1979). Momentary excessive construction on the basis for proadaptation. *Evolution*, 33, 227–33.
Garland, T. (1983). The relation between maximum running speed and body mass in terrestrial animals. *Journal of Zoology (Lond.)*, 199, 157–90.
Garland, T., Geiser, F. & Baudinette, R. V. (1988). Comparative performance of marsupial and placental mammals. *Journal of Zoology (Lond.)*, 215, 505–22.
Gauthier-Pilters, C. & Dagg, A. I. (1981). *The Camel: Evolution, Ecology, Behaviour and Relationship to Man*. University of Chicago Press.
Goldspink, G. (1992). The brains behind the brawn. *New Scientist*, 15th August, 33–7.
Gunn, H. M. (1989). Heart weight and running ability. *Journal of Anatomy*, 167, 225–33.
Harrison, D. F. N. & Denny, S. P. (1985). Possible influence of laryngeal and tracheal size on the running speed of mammals. *Acta Otolaryngologica (Stock.)*, 99, 229–35.
Harrison, D. F. N. & Denny, S. P. (1989). Symmorphosis in relation to the relative dimensions of the glottic and tracheal area in the mammalian larynx. *Journal of Laryngology and Otology*, 103, 1053–6.
Heath, D. & Williams, D. R. (1979). *Life at High Altitudes*. London, Edward Arnold.
Hinds, D. S. & Calder, W. A. (1971). Tracheal dead space in the respiration of birds. *Evolution*, 25, 429–40.
Hock, R. J. (1970). The physiology of high altitude. In *Vertebrate Structures and Functions*, p. 195–204. San Francisco, W. H. Freeman & Co.
Hoyt, B. & Taylor, C. R. (1981). Gait and the energetics of locomotion in horses. *Nature*, 292, 239–40.
Huxley, J. S. (1932). *Problems of Relative Growth*. London, Methuen.
Jones, J. D. (1972). *Comparative Physiology and Respiration*. London, Edward Arnold.

Ki Hong, S. & Rahn, H. (1974). The diving women of Korea and Japan. In *Vertebrates and Functions*, pp. 185–94. San Francisco, W. H. Freeman & Co.
Lighthill, M. J. (1977). Introduction to the scaling of aerial locomotion. In *Scale Effects in Animal Locomotion*, ed. T. J. Pedley, p. 365–484. London, Academic Press.
Martin, R. D. (1990). *Primate Origins and Evolution*. London, Chapman and Hall.
Maughan, R. (1992). Success on a plate. *New Scientist*, 25th July, 36–40.
McNeill Alexander, R. (1982). *Locomotion in Animals*. Glasgow, Blackie.
McNeill Alexander, R. (1983). *Animal Mechanics*. London, Blackwell Scientific Publications.
McNeill Alexander, R. (1988). *Elastic Mechanisms in Animal Movement*. Cambridge University Press.
McNeill Alexander, R. & Jayes, A. S. (1983). A dynamic similarity hypothesis for the gaits of quadrupedal mammals. *Journal of Zoological Society (Lond.)*, **201**, 135–52.
McNeill Alexander, R., Langman, V. A. & Jayes, A. S. (1977). Fast locomotion of some African Ungulates. *Journal of Zoology (Lond.)*, **183**, 291–300.
McNeill Alexander, R. & Pond, C. (1992). Locomotion and bone strength of the white rhino. *Journal of Zoology (Lond.)*, **227**, 63–9.
Napier, J. (1973). The antiquity of human walking. In *Vertebrates Structures and Functions*, p. 48–58. San Francisco, W. H. Freeman.
Napier, J. R. & Napier, P. H. (1985). *The Natural History of Primates*. London, British Museum (Natural History).
Negus, V. E. (1947). *Comparative Anatomy of the Larynx*. London, Heinemann.
Pond, C. (1989). Bearing up in the Artic. *New Scientist*, 4th February, 56.
Pye, J. D. (1968). Animal sonar in air. *Ultrasonics*, **6**, 32–8.
Schmidt-Nielson, K. (1983). *Animal Physiology*. 3rd edn. Cambridge University Press.
Seeherman, H. J., Taylor, C. R. & Malory, G. M. O. (1981). Design of the mammalian respiratory system. *Respiratory Physiology*, **44**, 11–23.
Sharp, C. & Parry-Billings, M. (1992). Can exercise damage your health? *New Scientist*, 15th August, 33–7.
Taylor, C. R. (1978). The energetics of terrestrial locomotion and body size in vertebrates. In *Scale Effects in Animal Locomotion*, ed. T. J. Pedley, pp. 127–41. London, Academic Press.
Taylor, C. R. & Weibel, E. R. (1981). Design of the mammalian respiratory system. *Respiratory Physiology*, **44**, 1–10.
Turnbull, A. (1992). Making all the right moves. *New Scientist*, 25th July, 23–7.
Vaughan, T. A. (1985). *Mammology*. 3rd edn, pp. 108–34. New York, Saunders College Publications.
Vines, G. (1987). Science goes to the dogs. *New Scientist*, 29th October, 45–7.
Vines, G. (1992). Winning streak for sheiks. *New Scientist*, 19th December, 32–5.
Yamashivo, S. M. & Grodin, F. S. (1973). Respiratory cycle optimisation in exercise. *Journal of Applied Physiology*, **35**, 522–5.

5.3 Vocalization

Aitchison, J. (1978). *The Articulate Mammal*. London, Hutchinson.
Below, P. (1551). *L'Histoire Naturelle des Estanges Poissins Marins*. Paris, R. Chaudiere.
Bright, N. (1984). *Animal Language*. London, British Broadcasting Association.

Caldwell, M. C. (1968). Vocalization of naive captive dolphins in small groups. *Science*, 159, 1121–3.

Denny, S. P. (1976). Comparative antomy of the larynx. In *Scientific Basis of Otolaryngology*, ed. R. Hinchcliffe & D. N. F. Harrison, p. 536. London, Heinemann.

Evans, W. E. (1973). Echolocation by marine delphinids and one species of fresh water dolphin. *Journal Acoustic Society of America*, 54, 191–9.

Ferrein, A. (1741). De la formation de la voix de l'homme. *History of the Royal Academy*. London, Royal Academy.

Fossy, D. (1972). Vocalizations of the mountain gorilla. *Animal Behaviour*, 20, 36–53.

Gaskin, D. E. (1982). *The Ecology of Whales and Dolphins*. London, Heinemann.

Gould, W. J. & Korouin, S. (1994). Laboratory Advances in voice measurements. *Journal of Acoustical Society of America*, 80, 391–401.

Haimoff, E. H. (1984). Acoustic and organizational features of gibbon songs. In *The Lesser Apes*, ed. Preuschft, Chivers & Brockelman, p. 333–53. Edinburgh University Press.

Haji, T., Mori, K. & Omori, K. (1992). Experimental studies on the viscoelasticity of the vocal folds. *Acta Otolaryngologica (Stock.)*, 112, 151–9.

Hast, M. (1989). The larynx of roaring and non-roaring cats. *Journal of Anatomy*, 163, 117–21.

Hemmer, H. (1981). Die Evolution der Pantherkatzen. *Paleontology Zoologica*, 55, 109–16.

Hirano, M. (1981). *Clinical Examination of the Voice*. New York, Springer-Verlag.

Hollien, H. & Shipp, T. (1972). Speaking fundamental frequency and chronological age in males. *Journal of Speech and Hearing Research*, 15, 155–9.

Honjo, I. & Isshiki, P. (1980). Laryngoscopic and voice characteristics of aged persons. *Archives of Otolaryngology (Stock.)*, 106, 149–50.

Hunsaker, D. & Hahn, T. C. (1965). Vocalization of the South American tapir. *Animal Behaviour*, 13, 69–74.

Jakobson, R. (1940). Child language, aphasia and phonological universals. The Hague, Mouton. (Translated by A. R. Keller.)

Kiley, M. (1972). The vocalization of ungulates, their causation and function. *Zoology Tierpsyche*, 31, 171–222.

Koyama, T., Kawasaki, M. & Ogura, J. H. (1969). Mechanics of voice production. 1. Regulation of vocal intensity. *Laryngoscope*, 56, 337–54.

Lieberman, P. (1991). *The Evolution of Uniquely Human Speech, Thought and Selfless Behaviour*. Harvard University Press.

Lieberman, P., Harris, K. S. & Wolff, P. (1971). New born infant cry and non-human vocalization. *Journal of Speech and Hearing Research*, 14, 718–27.

Messel, R. (1993). Barking dogs are stuck in adolescence. *New Scientist*, 27th February, 9.

Mitchell, A. W. (1986). *The Enchanted Canopy*, p. 112. London, Collins.

Muller, J. (1837). *Von der Stimmer und Sprache, Hundbuch der Physiol des Menschen*, vol 2, book 4, pp. 133–245.

Mysak, E. D. (1959). Pitch and duration characteristics of older males. *Journal of Speech and Hearing Research*, 2, 46–54.

Owen, R. (1834). On the Anatomy of the Cheetah. *Transactions of the Zoological Society of London*, 1, 129–36.

Passingham, R. (1982). *The Human Primate*, pp. 194–6. Oxford, W. H. Freeman & Co.

Peters, G. (1984). A special type of vocalization in felids. *Acta Zoologica Fennica*, 171, 89–92.

Peters, G. (1985). A comparative survey of vocalization in the giant panda. *Proceedings of the International Symposium on the Giant Panda*, pp. 197–208. Berlin, Bongo.

Pocock, R. I. (1916). On the hyoid apparatus of the lion and related species of Felidae. *The Annals and Magazine of Natural History*, 8, 222–9.

Pola, Y. & Snowdon, C. T. (1975). Vocalizations of pygmy marmosets. *Animal Behaviour*, 23, 826–42.

Pressman, S. S. & Keleman, G. (1955). Physiology of the larynx. *Physiology Reviews*, 35, 306–9.

Punt, N. (1979). *The Singers and Actors Throat*. Heinemann, London.

Reidenberg, J. S. & Laitman, J. T. (1988). Existance of vocal folds in the larynx of Odontoceti (toothed whales). *Anatomical Record*, 221, 884–91.

Remmers, J. E. & Gautier, H. (1972). Neural and mechanical mechanisms of feline purring. *Respiratory Physiology*, 16, 351–61.

Sales, G. D. (1970). Ultrasonic communication in rodents. *Nature*, 191, 531–55.

Schaller, G. B., Jinchu, H. & Wenshi, P. (1985). *The Giant Pandas of Wolong*. University of Chicago Press.

Seyfarth, R. M., Cheney, D. L. & Marley, P. (1980). Vervet monkeys alarm calls: semantic communication in a free ranging primate. *Animal Behaviour*, 28, 1070–94.

Shanks, D. (1993), Breaking Chomsky's rules. *New Scientist*, 30th January, 26–30.

Sissom, D. E. F. & Rice, D. A. (1991). How cats purr. *Journal of Zoology* (Lond.), 223, 67–78.

Small, M. F. (1994) A chimp with an accent. *New Scientist*, 4th June, 33–7.

Sonesson, B. (1960). On the anatomy and vibratratory pattern of the human vocal cords. *Acta Otolaryngologica (Stock.)*, Suppl. 156.

Spallanzani, L. (1932). *Opere di Lazzara Spallanzanic tolzi*. Milan.

Thumfart, W. F. (1988). From larynx to vocal ability. *Acta Otolaryngologica (Stock.)*, 105, 425–31.

Titze, R. (1993). Current topics in voice production mechanisms. *Acta Otolaryngologica (Stock.)*, 113, 421–7.

Titze, R. (1994). Towards standards in acoustic analysis of voice. *Journal of Voice*, 8, 1–7.

Titze, R. & Talkin, D. T. (1979). A theoretical study of the effects of various laryngeal configurations on the acoustics of phonation. *Journal of the Acoustical Society of America*, 66, 60–74.

Van Gulik, R. H. (1967). *The Gibbon in China: An Essay on Chinese Animal Lore*. The Netherlands, E. S. Brill.

Vaughan, T. A. (1986). *Mammology*, 3rd edn., pp. 494–506. Philadelphia, Saunders College Publishing.

Watts, C. H. S. (1975). Vocalization in Australian hopping mice (*Notomys*). *Journal of Zoology (Lond.)*, 177, 247–63.

Watts, C. H. S. (1980). Vocalization in nine species of rat. *Journal of Zoology (Lond.)*, 191, 531–55.

Wood, S. (1986). The acoustic significance of tongue, lip and larynx maneuvers in rounded palatal vowels. *Journal of the Acoustical Society of America*, 80, 391–401.

6 Evolutionary concepts

For much of the twentieth century biologists resisted the concept that humans and other animals were part of an evolutionary continuum. They accepted that humans evolved from an ape-like ancestor but regarded tool use, culture and language as uniquely human. More recently, the size and nature of the 'hiatus' between humans and non-humans, specifically the great apes, has exercised the minds of workers in many disciplines (Gibson & Ingold, 1993). Research has shown that other primates show evidence of cognitive skills such as tool use or language; this concept is, however, resisted by those who believe that such attributes only occurred with the advent of *Homo sapiens*. The belief that 'humanness' appeared *de novo*, fully developed within the modern human brain and separating 'Man' from his ancestors, ignores the palaeoanthropological evidence but fuels philosophical discussion!

Much of the evidence for the origin of modern humans rests on the interpretation of fragments of bone, enigmatic archaeological discoveries or DNA gels. Whilst these provide intriguing signposts to the past and our relationships to the great apes, they do nothing to explain 'Man's' bipedal stance, large brain and other human characteristics. The theory of 'punctuated equilibrium' suggests that all species undergo spurts of evolutionary change. These may be initiated by environmental changes – Leakey & Lewin (1992) proposing that global cooling was the driving force behind the evolution of humans. However, the chronological history of human evolution remains in dispute, although becoming clearer with discoveries of larger numbers of intact fossils, more accurate dating and advances in molecular anthropology.

Palaeoanthropological evidence for the emergence of human striding bipedalism is limited by availability of fossil records before 4 million years ago, the oldest known evidence being a footprint trail from the Laetoli site in Tanzania dated 3.6 million years old (Leakey & Hay, 1979). Fossil foot bones attributed to *H. habilis* found in the Olduvia Gorge, suggested a configuration somewhere between the great apes and man and were dated about 1.8 million years old. The timing of the inception of human bipedal locomotion must await further fossil discoveries although clearly, it owed much to previous adaptations to arboreal life (Martin, 1993).

Early hominids such as *Australopithecus afarensis* had small brains but hip and knee joints suitable for bipedal locomotion. Palaeoanthropologists recognize that this was the most important step in the evolution of hominids, but disagree as to why this form of locomotion was adopted. Freedom of hands for tool usage or weapon handling was certainly a secondary benefit, although neither of these abilities appeared until more than two million years after *Australopithecus*. Theories that a need for long distance travel, food carrying or collecting fruit iniated bipedalism has limited support because the anthropological evidence suggests that early hominids were poorly designed for an upright stance. Fragmentation of the forest cover from global changes and the consequent dispersal of food have also been suggested as a stimulus to terrestrial locomotion. Bipedal apes would be more successful in food foraging, leading to the evolution about two million years ago of *H. erectus*, a hominid well adapted for bipedal locomotion and a mixed diet. These were larger than previous hominids, meat eaters and required expansion of their geographical range. The earliest known stone tools are dated about 2.5 million years old with more sophisticated teardrop-shaped hand axes (named Acheulian after St Acheul where the first axes were found) appearing one million years later. Doubt remains as to whether *H. erectus* remained in Africa for the next 800,000 years before expanding into Asia, or left much earlier but failed to develop tool 'technology' in all of its new sites (Lewin, 1994).

Homo erectus, however, showed definitive progression from an ape-like past to what was to be a human-like future, being replaced about 130,000 years ago by hominids who later became 'early archaic' humans. Called Neanderthals after the Neander Valley in which bones were first discovered, they existed for 35,000 years before evolving (or being superseded by) our immediate predecessors. Some palaeoanthropological evidence suggests similarities with modern skeletal remains: large brains and evidence of cultural behaviour. Widely distributed throughout Europe and Southwest Asia, controversy remains as to whether they were eventually replaced by *H. erectus* coming from Africa or evolved into modern Europeans.

Proponents of the theory that 'Man' developed solely in the continent of Africa argue that the principal human races: Negroids, Caucasoids, Mongoloids, Australian aboriginal people and Southern African Bushman, only began to evolve at this time. The opposing school believe that anatomically, modern humans evolved from archaic hominids simultaneously in different parts of the world. During this period of evolution, racial characteristics of

H. sapiens emerged, although clarification of this contentious issue awaits analysis of thousands of DNA samples from hundreds of different ethnic groups. Moreover, human groups cannot be defined solely by genetic characteristics, even when they vary from group to group in a distinctive manner. It is anticipated that data relating to genetic differences based on variations in mitochondrial DNA will confirm the hypothesis that all modern humans originated from Africa, between 100,000 and 200,000 years ago. Information about genetic diversity could also give insight into population movements since the evolution of *H. sapiens* (Lewin, 1993a).

Evolution of speech and descent of the larynx

Historical linguists when tracing language origins compare words that are similar in different languages, then trace them back to a common ancestor. There is evidence from the Classical Sanskrit of India that most European, Indian and Iranian languages are related to one common language spoken 8000 years ago – at the time of domestication of the horse and invention of the wheel (Stevenson, 1983). Linguistic research suggests that traces of languages spoken over 50,000 years ago can be detected by grammatical analysis of modern spoken languages. Word sounds change rapidly, and research has concentrated on the published grammars of more than 300 modern language groups. By comparing New World languages it has been estimated that the first settlers in America arrived around 35,000 years ago speaking at least 10 different linguistic groups (Kleiner, 1994). Although the more orthodox school of linguists doubt the validity of 'retro-extrapolation' of historical linguistics, the data do suggest that by 50,000 years ago human language was already diverse. Language, although natural to modern humans is prone to corruption, decay and ultimately extinction. Its survival is ultimately related to the fate of individuals and like them, must adapt to changing circumstances.

Speculation on the evolution of human language must therefore consider two questions. Did it evolve as part of a continuous spectrum of brain development shared with apes or did it emerge as a fundamental difference in the evolution of human and primate behaviour? Those who believe language to be part of a continuous spectrum of evolving behaviour accept that its evolution must date to the time of our common ancestors with apes, over 5 million years ago. Since that time, apes have developed along different anthropological lines and could not therefore be expected to have primitive language ability. Human language and its associated neural structures are so complex that their evolution must have occurred gradually over a much longer period of time than is possible from the origins of modern *Homo*, 2–3 million years ago. Although large brains are not directly related to intellectual ability, the human brain is three times that of the chimpanzee when correlated with differences in body size. Fossil brain casts are the only source of anatomical data of brain organization and development. From inferred body weight, cranial capacities have been estimated by Martin (1983) as 442 cm³ for *Australopithecus africanus*, 642 cm³ for *Homo habilis*, 941 cm³ for *H. erectus* and 1230 cm³ for *H. sapiens*. These confirm the considerable increase in cerebral capacity that occurred during the evolution of hominids, as well as providing information relevant to development of the neocortex. Such data, however, provide little information regarding the complex neural pathways and connections required for the production and comprehension of human language. There is evidence that major evolutionary changes, such as the origin and development of modern mammals, were accompanied by an increase in brain size. This is not necessarily associated with an increase in intelligence but as Leakey & Lewin (1992) have argued, was the result of modifications of existing or the development of new sensory channels. Extrapolation of this argument leads to the conclusion that a bigger more complex brain was needed for the development of human speech, with its range of sounds, syntax and grammatical structure (Martin, 1990).

Vocal communication exists in most mammals and is relatively complex in primates, whose brains are clearly competent to respond to species-specific vocalizations. Whether this should be equated with human language is of less interest to the comparative laryngologist than to practitioners of psycholinguistics. What has attracted considerable attention, however, is whether the study of laryngeal morphology can reveal at what stage early hominids were functionally equipped to talk. Previous chapters have demonstrated considerable similarity between primate larynges, with the exception of the large ventricular air sacs present in the great apes. Whether these have a respiratory function is doubtful and they appear to play only a minimal role in increasing the resonance of primate vocalization.

Humans, however, are the only mammals who in adulthood have a larynx set low in the neck. This results in a large pharynx and a tongue that is now partly pharyngeal causing the epiglottis and palate to separate. Despite loosing the protective value of an intranarial epiglottis, the benefit of this descent has been an ability to produce a range of sounds unique amongst mammals. Reduction in epiglottic-palatal contact is apparent in many primates. In dissected specimens of the great apes and gibbons there is distinct separation, although this may not reflect the true position during life. In these primates, however, the pharynx is small and the long thin tongue entirely intraoral. Lowering of the human larynx is considered to be the end point of a trend visible throughout the primates; together with appropriate cerebral development it made human speech possible, giving man a unique advantage over all other animals.

Could Neanderthals speak?

In his search for a morphological explanation for the failure of non-human primates to speak, Crelin (1987) compared the skull base of the adult chimpanzee with that of a one-year-old human. Both have larynges situated high in the neck and wide intervals between the vomer and synostosis of what was the spheno-occipital synchondrosis. During development of the skull base this is the area where most growth occurs, becoming a bony synostosis at puberty. Since the roof of the oral cavity and nasopharynx are directly related to the skull base, Crelin believed that study of this region in primates and hominids would indicate the timing of laryngeal descent.

As a child develops the distance between the vomer bone and spheno-occipital synchondrosis gradually decreases, and results in a similar reduction of the distance between hard palate and foramen magnum. This area gradually forms a deep concavity and by 6 years of age the vomer has reached the basilar part of the occipital bone. This only occurs in the adult human and is associated with a larynx low in the neck. Laitman et al. (1978) carried out statistical analysis of skull base measurements of chimpanzees. They concluded that the distance between the back of the hard palate and vertebral column was directly associated with flexure of the skull base and correlated with laryngeal position and tongue shape. Both of these are essential components in the production of speech. Crelin (1987) showed from his studies on latex rubber reconstructions of the vocal tracts of adult chimpanzees and human infants, that both were limited in their range of vocalization. He concluded that all non-human primates were anatomically limited in their ability to produce the full range of sounds necessary for human speech, irrespective of the presence of adequate brain capacity and neuromuscular co-ordination.

The major achievement of modern humans was surely the acquisition of a fully articulated spoken language. This crucial advance was not just an ability to produce the 50 or more sounds on which human speech is formulated, but the brain to co-ordinate and utilize them for those myriad functions that separate *Homo sapiens* from the great apes. Fossil records show that Australopithecines with their small brains had ape-like basicraniums and attention has focused on the largest group of archaic hominids, the Neanderthals. Bipedal and with large brains they lived between 125,000 and 35,000 years ago, eventually being replaced by modern 'Man'. During this lengthy existence modification of the supralaryngeal tract must have occurred, although lack of adequate fossil evidence has restricted accurate assessment of the period when a fully flexed basicranium can be detected. Crelin (1987) studied casts made from a variety of hominid skulls ranging from *Australopithecus africanus* (Mrs Ples, with brain size of 482 cm^3), *H. habilis* (brain size 700 cm^3), *H. erectus* (600–1,000 cm^3) to archaic *H. sapiens* (ranging from 200,000 to 45,000 years ago). The cranial cavities of the latter varied around 1,300 cm^3, but many of the fossils were incomplete or damaged, making reconstruction and subsequent evaluation of the skull base imprecise. Only three skulls were considered to be classical Neanderthals rather than 'recent' archaic *H. sapiens*. The most contentious proved to be that found in a cave at La Chapelle-aux-Saints in France, dated about 50,000 years old. This skull was considered to have been interred at a ritual burial, and Crelin's reconstruction suggested that there was an intranarial epiglottis and a completely intra-oral tongue. Lieberman *et al.* (1972) concluded from this assessment that the airway of Neanderthals was similar to that of non-human primates but that 'they probably had the ability to produce nasalized versions of all sounds of human speech'. The skeleton of this 'old man' of La Chapelle was deformed and reconstruction possibly distorted the true degree of basicranium flexion. This emphasizes the difficulty and need for caution when attempting to rationalize on the basis of limited anatomical evidence. In 1983, a 60,000-year-old Neanderthal skeleton was discovered in a cave on Mount Carmel, Israel. Within the mandible lay a hyoid bone, the first to be found in a fossilized ancestral human. This caused considerable interest, with Arensburgh *et al.* (1989) suggesting that it provided proof of human speech. This bone, however, bore only partial resemblance to modern hyoids and provided little information regarding supralaryngeal structure or laryngeal position (Marshall, 1989).

All mammals have hyoid bones that are usually ossified and provide support for the tongue and a range of muscle and fibrous attachments to the skull base, mandible and larynx. The size and marking of the hyoid bone helps in pinpointing the tongue position, pharyngeal size and possibly laryngeal position. To associate their morphology with vocal function requires more evidence, which at present is unavailable. Detachment of the hyoid as part of operations to release the trachea in patients with stenosis does not affect their speech.

The most acceptable testimony to the intellectual, and by association, vocal abilities of Neanderthals is best sought in their cultural attainments. Concern for the dead expressed by some form of ritual burial, indicates both self-awareness and consciousness. Association with stone tools, floral tributes or body alignment suggests the possibility of some articulate communication, although their large brains probably did not contain the fully developed neurological pathways required for modern speech. Archaic hominids existed for about 50,000 years during which time the gradual descent of the larynx with concomitant development of appropriate neurological pathways could be expected as part of an evolutionary process, which ultimately was to result in modern *H. sapiens*. The palaeontological and archaeological records of archaic *Homo* vary widely between Africa, Asia and Europe, reflecting both structural differences and limitations in fossil evidence (Lewin, 1993b). If, however, development of a fully propositional language is related to a larynx low in the neck, a tongue which is partially pharyngeal and a brain programmed to deal adequately with the flow of auditory information, speculation as to the timing of this uniquely human asset will continue until more reliable fossil evidence is available.

Negus in 1929 in his *Mechanism of the Larynx* emphasized that the primary function of this organ was airway protection. Descent of the adult larynx from its relatively secure position high in the neck, has increased the vulnerability of the respiratory tract to overspill (Lieberman, 1991). This is the price that evolution has demanded for the acquisition of speech, a facility which has placed modern. *H. sapiens* at the top of the Primate Tree, where 'he' perches precariously.

New data has been reported in this book, which together with existing information has been examined with the expectation that it would clarify the role the larynx plays in controlling mammalian behaviour. Lack of detailed information relating to mammalian physiology and comparative morphology, enevitably restricts clarification of many normal physiological functions. By considering each laryngeal function individually it was hoped at best, to clarify and at worst, to highlight those areas where additional information is needed. Collection of adequate numbers of the many mammalian species presents considerable problems. Without this material, however, further progress in our understanding of the complexities and subilities of laryngeal function will be impossible.

'*Felix qui potuit rerum cognoscere causas*', which translates approximately as 'Fortunate is he who has been able to learn the cause of things',

> What resources make change
> In nature and man,
> That the Universe could happen
> Under any circumstances,
> And that this mystery
> Is the total progenitor of all energy
> And all its happenings
> In an endless replication

Of the success of chance,
Inside the secret intelligence that manipulates the source,
Which remains unknown – even to its chemicals
In a geometry of evolution
That survives or is lost,
And on top of this
Comes a brain machinery
Employed in a thought process
From these enigmas
That has invaded the source –
But doesn't know its destiny after this religious myth.

John Conley (1992), Laryngologist *par excellence.*

References

Arensburgh, B., Tiller, S. & Vandermeersch, H. (1989). A middle paleolithic hyoid bone. *Nature*, 338, 758–60.

Crelin, E. S. (1987). *The Human Vocal Tract.* New York, Vantage Press.

Gibson, K. R. & Ingold, T. (1993). *Tools, Language and Cognition in Human Evolution.* Cambridge University Press.

Kleiner, K. (1994). Echoes of ancient Africa in our speech. *New Scientist*, 23rd April, 10.

Laitman, J. T., Heimbuch, R. C. & Crelin, E. S. (1978). Developmental change in a basicranial line and its relationship to the upper respiratory system in living primates. *American Journal of Anatomy*, 152, 467–82.

Leakey, M. D. & Hay, R. L. (1979). Pliocene footprints in the laetotic beds at Laetotic, Tanzania. *Nature*, 278, 317–23.

Leakey, R. & Lewin, R. (1992). *Origins Reconsidered.* London, Little Brown & Co.

Lewin, R. (1993a). Genes from a disappearing world. *New Scientist*, 29th May, 25–9.

Lewin, R. (1993b). *The Origin of Modern Humans.* New York, Scientific American Library.

Lewin, R. (1994). Human origins: the challenge of Java's skulls. *New Scientist*, 7th May, 36–40.

Lieberman, P., Crelin, E. S. & Klatt, D. H. (1972). Phonetic ability, related anatomy of the newborn, adult human, Neanderthal man and chimpanzee. *American Anthropologist*, 74, 287–307.

Lieberman, P. (1991). *Uniquely Human.* Harvard University Press.

Marshall, J. C. (1989). The descent of the larynx. *Nature*, 338, 702–3.

Martin, R. D. (1983). *Human Brain Evolution in a Ecological Context.* New York, American Museum of Natural History.

Martin, R. D. (1990). *Primate Origins and Evolution.* London, Chapman & Hall.

Negus, V. (1929). *The Mechanism of the Larynx.* London, Heinemann.

Stevenson, V. (1983). *Words*, pp. 10–17. London, Eddison-Sadd Editions.

Appendices

Breakdown by species and age for 1410 mammals examined

ORDER MONOTREMATA

Species	Common name	Baby	Junior	Adult	Total
Zaglossus bruisnii	Long-nosed echidna	0	0	2	2

ORDER MARSUPIALIA

Species	Common name	Baby	Junior	Adult	Total
Monodelphis domestica	Short-tailed opossum	2		6	8
Metachirus nudicaudatus	Four-eyed opossum			5	5
Didelphis virginiana	Virginian opossum	1		6	7
Dasyurus byrnei	Kowari			4	4
Sarcophilus harrisii	Tasmanian devil			4	4
Sminthopsis crassicaudata	Fat-tailed dunnart			1	1
Perameles nasuta	Lesser mouse-tailed bat		1	2	3
Perameles gunnii	Barred bandicoot			1	1
Echymipera kalubu	Spiny bandicoot			2	2
Trichosurus vulpecula	Brushtail possum			3	3
Phalanger orientalis	Grey cuscus		2	2	4
Petaurus breviceps	Sugar glider			3	3
Dactylopsila trivirgata	Striped possum			4	4
Potorous tridactylus	Long-nosed potoroo			3	3
Macropus agilis	River wallaby		1	2	3
Macropus parma	Parma wallaby	4	4	12	20
Macropus robustus	Wallaroo	1	3	9	13
Macropus rufus	Red kangaroo		2	6	8
Macropus rufugriseus	Bennett's wallaby	2	3	25	30
Phascolarctos cinereus	Koala		1	3	4

ORDER MARSUPIALIA (*cont.*)

Species	Common name	Baby	Junior	Adult	Total
Vombatus ursinus	Common wombat	1		5	6
		11	17	108	136

ORDER EDENTATA

Species	Common name	Baby	Junior	Adult	Total
Myrmecophaga tridactyla	Giant anteater			2	2
Bradypus variegatus	Three-toed sloth			1	1
Choloepus didactylus	Linne's two-toed sloth			1	1
Chaetophractus villosus	Large hairy armadillo			1	1
Dasypus novemcinctus	Nine-banded armadillo			4	4
		0	0	9	9

ORDER INSECTIVORA

Species	Common name	Baby	Junior	Adult	Total
Echinops telfairi	Hedgehog tenrec	1		3	4
Erinaceus europaeus	European hedgehog			6	6
Sorex araneus	Common shrew	4		8	12
Talpa europaea	Common mole	1		4	5
		6	0	21	27

ORDER CHIROPTERA

Species	Common name	Baby	Junior	Adult	Total
Rousettus aegyptiacus	Egyptian rousette			3	3
Pteropus giganteus	Indian flying fox		2	5	7
Epomops franqueti	'Singing' fruit bat			3	3
Micropteropus pusillus	Dwarf fruit bat			2	2
Rhinopoma hardwickii	Lesser mouse-tailed bat			4	4
Saccopteryx bilineata	Sac-winged bat			5	5
Nycteris macrotis	Slit-faced bat			4	4

ORDER CHIROPTERA (*cont.*)

Species	Common name	Baby	Junior	Adult	Total
Nycteris thebaica	Egyptian slit-faced			1	1
Cardioderma cor	Heart-nosed bat			2	2
Lavia frons	Yellow-winged bat			2	2
Rhinolophus clivosus	Geoffroy's horseshoe			1	1
Rhinolophus ferrumequinum	Greater horseshoe			8	8
Rhinolophus hipposideros	Lesser horseshoe			6	6
Hipposideros caffer	Leaf-nosed bat			2	2
Noctilio leporinus	Fisherman bat			6	6
Phyllodia parnelli	Moustached bat			1	1
Phyllostomus hastatus	Spear-nosed bat			3	3
Glossophaga soricina	Long-tongued bat			2	2
Desmodus rotundus	Common vampire bat		2	6	8
Natalus tumidirostris	Funnel-eared bat			1	1
Pipistrellus pipistrellus	Common pipistrelle			8	8
Molussus molussus	Mastiff bat			4	4
		0	4	79	83

ORDER PRIMATES

Species	Common name	Baby	Junior	Adult	Total
Microcebus murinus	Lesser mouse-lemur			3	3
Lemur catta	Ring-tailed lemur	2	3	8	13
Petterus fulvus	Brown lemur	1		2	3
Varecia variegata	Ruffed lemur		2	6	8
Loris tardigradus	Slender loris		1	3	4
Nycticebus coucang	Slow loris			4	4
Otelemur crassicaudatus	Greater bushbaby		1	4	5
Galago senegalensis	Lesser bushbaby			1	1
Callithrix argentata	Silvery marmoset			1	1
Callithrix jacchus	Common marmoset	22	12	58	92
Cebuella pygmaea	Pigmy marmoset			4	4
Saguinus imperator	Emperor tamarin			1	1
Saguinus labiatus	White-lipped tamarin	3	1	2	6
Saguinus oedipus	Cotton-top tamarin	3	1	5	9
Leontopithecus rosalia	Golden lion tamarin			5	5
Cebus apella	Brown capuchin	1		4	5
Saimiri sciureus	Common squirrel monkey			3	3
Aotus trivirgatus	Northern night monkey			1	1
Pithecia pithecia	White-faced saki	1		2	3

ORDER PRIMATES (*cont.*)

Species	Common name	Baby	Junior	Adult	Total
Alouatta caraya	Black howler			1	1
Ateles paniscus	Black spider monkey	1		2	4
Macaca irus	Crab-eating macaque			2	2
Macaca nemestrina	Pig-tailed macaque	2		3	5
Macaca silenus	Lion-tailed macaque	1		2	3
Macaca sylvanus	Barbary ape	1	2	2	5
Cercocebus atys	Sotty mangabey	3	3	5	11
Papio hamadryas	Hamadryas baboon			1	1
Mandrillus sphinx	Mandril			3	3
Theropithecus gelada	Gelada		2	2	4
Cercopithecus neglectus	De Brazza's monkey	1		4	5
Cercopithecus pygerythrus	Green monkey			2	2
Colobus satanus	Black colobus			2	2
Tupaia belangeri	Northern tree shrew	1		3	4
Tupaia tana	Large tree shrew			2	2
Hylobates lar	Common gibbon	3	1	6	10
Hylobates syndactylus	Siamang			1	1
Pongo pygmaeus	Orang-utan	3	3	3	9
Pan troglodytes	Chimpanzee	1	2	5	8
Gorilla gorilla	Gorilla	2		3	5
		52	34	172	258

ORDER CARNIVORA

Species	Common name	Baby	Junior	Adult	Total
Canis aureus	Golden jackal			1	1
Canis latrans	Coyote			2	2
Canis lupus	Wolf	2	2	7	11
Vulpes vulpes	Red fox	1		3	4
Vulpes zerda	Fennec fox			2	2
Urocyon cinereoargenteus	Grey fox		1		1
Lycaon pictus	Hunting dog	2		6	8
Selenarctos thibetanus	Asiatic black bear			1	1
Ursus americanus	American black bear		1	2	3
Ursus arctos	Brown bear		2	1	3
Thalarctos maritimus	Polar bear			1	1
Melursus ursinus	Sloth bear			1	1
Ailuropoda melanoleuca	Giant panda			1	1
Procyon lotor	Common racoon	2	2	3	7
Nasua nasua	Coati	1		4	5

ORDER CARNIVORA (*cont.*)

Species	Common name	Baby	Junior	Adult	Total
Potos flavus	Kinkajou			1	1
Ailurus fulgens	Red panda			3	3
Arctonyx collaris	Hog badger			2	2
Meles meles	European badger	2	2	18	22
Lutra canadensis	Canadian otter			3	3
Lutra lutra	European otter	1	1	2	4
Enhydra lutris	Sea otter			1	1
Viverra civetta	African civet			2	2
Viverra zibetha	Indian civet			1	1
Herpestes fuscus	Indian grey mongoose			2	2
Ichneumia albicauda	White-tailed mongoose			1	1
Suricata suricatta	Meerkat		1	2	3
Proteles cristatus	Aardwolf			1	1
Hyaena brunnea	Brown hyaena			1	1
Felis caracal	Caracal			2	2
Felis concolor	Puma			3	3
Felis pardalis	Ocelot			1	1
Felis serval	Serval			2	2
Felis wiedii	Margay		1	2	3
Lynx lynx	Canadian lynx	1		3	4
Lynx rufus	Bobcat		1	1	2
Panthera leo	Lion	4	2	8	14
Panthera onca	Jaguar	1	2	4	7
Panthera pardus	Leopard	2	1	6	9
Panthera tigris	Tiger	1	1	4	6
Acinonyx jubatus	Cheetah	2		14	16
		22	20	125	167

ORDER PINNIPEDIA

Species	Common name	Baby	Junior	Adult	Total
Zalophus californianus	Californian sealion	3	1	4	8
Phoca vitulina	Common seal	2		5	7
Halichoerus grypus	Grey seal			3	3
		5	1	12	18

ORDER CETACEA

Species	Common name	Baby	Junior	Adult	Total
Delphinus delphis	Common dolphin	1			1
Tursiops truncatus	Bottlenose dolphin	3		1	4
Phocoena phocoena	Harbour porpoise			1	1
Balaenoptera musculus	Blue whale	1			1
Balaenoptera physalus	Fin whale	1			1
Megaptera novaeangliae	Humpback whale	1			1
		7	0	2	9

ORDER PROBOSCIDEA

Species	Common name	Baby	Junior	Adult	Total
Loxodonta africana	African elephant	1		3	4
Elephas maximus	Indian elephant			2	2
		1	0	5	6

ORDER PERISSODACTYLA

Species	Common name	Baby	Junior	Adult	Total
Equus burchelli	Common zebra	5	7	12	24
Equus grevyi	Grevy's zebra	1	2	7	10
Equus hemionus	Onager		2	8	10
Equus przewalski	Wild horse	1	1	6	8
Tapiris terrestris	Brazilian tapir	1		2	3
Ceratotherium simum	White rhino	1		3	4
Diceros bicornis	Black rhino			2	2
		9	12	40	61

ORDER HYRACOIDEA

Species	Common name	Baby	Junior	Adult	Total
Dendrohyrax arboreus	Tree hyrax			1	1
Heterohyrax brucei	Rock hyrax	1		4	5
Procavia capensis	Large-toothed rock hyrax			7	7
		1	0	12	13

ORDER ARTIODACTYLA

Species	Common name	Baby	Junior	Adult	Total
Sus scrofa	Wild boar	7	2	9	18
Phacochoerus aethiopicus	Wart hog	2		5	7
Tayassu tajacu	Collared peccary	2		12	14
Hippopotamus amphibius	Hippopotamus		1	4	5
Choeropsis liberiensis	Pigmy hippopotamus	1		2	3
Lama glama	Llama	1		4	5
Lama guanicoe	Guanaco			3	3
Lama pacos	Alpaca			1	1
Vicugna vicugna	Vicugna			1	1
Camelus bactrianus	Bactrian camel	1	2	13	16
Camelus dromedarius	Dromedary			4	4
Hyemoschus aquaticus	Water chevrotain			1	1
Hydropotes inermis	Chinese water deer	1		5	6
Muntiacus muntjak	Indian muntjac	1		9	10
Muntiacus reevesi	Chinese muntjac	1		5	6
Cervus axis	Spotted deer			4	4
Cervus dama	Fallow deer	4	2	11	17
Cervus duvaucelii	Swamp deer			4	4
Cervus elaphus	Red deer			2	2
Cervus nippon	Sika deer			3	3
Cervus porcinus	Hog deer	1		2	3
Cervus timorensis	Timor deer			4	4
Elaphurus davidianus	Pere David deer	2	2	8	12
Alces alces	Moose			2	2
Rangifer tarandus	Reindeer/caribou	5	2	12	19
Pudu pudu	Southern pudu	2		3	5
Capreolus capreolus	Western roe deer	3	1	6	10
Okapia johnstoni	Okapi	2		1	3
Giraffa camelopardalis	Giraffe	1		6	7
Tragelaphus angasii	Nyala	3		2	5

ORDER ARTIODACTYLA (*cont.*)

Species	Common name	Baby	Junior	Adult	Total
Tragelaphus derbianus	Eland			1	1
Tragelaphus eurycerus	Bongo			1	1
Tragelaphus spekii	Sitatunga			2	2
Tragelaphus strepsiceros	Greater Kudu	3	2	11	16
Boselaphus tragocamelus	Nilgai	1		7	8
Bubalus arnee	Indian buffalo			1	1
Bos guarus	Gaur	1		3	4
Bos mutus	Yak	1		6	7
Bos taurus	Aurochs			1	1
Synceros caffer	African buffalo	2		8	10
Bison bison	Bison	4	1	8	13
Cephalophus maxwellii	Maxwell's duiker			2	2
Kobus defassa	Waterbuck	2	2	11	15
Hippotragus equinus	Roan antelope	1	3	6	10
Hippotragus niger	Sable antelope	1		5	6
Oryx tao	Scimitar oryx	10	2	9	21
Oryx gazella	Gemsbok			4	4
Connochaetes taurinus	Blue wilderbeest	1	2	7	10
Damaliscus dorcas	Blesbok	2	1	6	9
Aepyceros melampus	Impala	2		2	4
Antilope cervicapra	Blackbuck	8	3	16	27
Antidorcas marsupialis	Springbok	1		4	5
Gazelle thomsonii	Thomson's gazelle	2		9	11
Ovibos moschatus	Musk ox	1		2	3
Capra falconeri	Markhor	6		5	11
Capra hircus	Wild goat	3	2	8	13
Ammotragus lervia	Barbary sheep	5	4	10	19
Ovis canadensis	American big-horn sheep	2	2	9	13
Ovis orientalis	Mouflon	6	2	9	17
		97	38	321	464

ORDER RODENTIA

Species	Common name	Baby	Junior	Adult	Total
Sciurus vulgaris	Red squirrel	2	2	5	9
Ratufa bicolor	Black giant squirrel			1	1
Tamiops swinhoei	Striped squirrel			1	1
Marmota marmota	Alpine marmot	1		3	4
Marmota monax	Woodchuck			3	3

274 Appendices

ORDER RODENTIA (*cont.*)

Species	Common name	Baby	Junior	Adult	Total
Cynomys ludovicianus	Black-tailed prairie dog			2	2
Tamias striatus	Eastern chipmunk	1		5	6
Glaucomys sabrinus	Northern flying squirrel			1	1
Geomys bursarius	Plains gopher	2		4	6
Castor canadensis	American beaver			2	2
Castor fiber	Eurasian beaver		2	5	7
Pedestes capensis	Springhaas	1		2	3
Lemniscomys striatus	Striped grass rat			1	1
Mesocricetus auratus	Golden hamster	4	4	6	14
Phloeomys cumingi	Philippine cloud rat			1	1
Glis glis	Edible dormouse			2	2
Dicrostonyx torquatus	Artic lemming			1	1
Arvicola terrestris	European water vole			3	3
Jaculus jaculus	Egyptian jerboa			3	3
Hystrix cristata	Crested porcupine			2	2
Atherurus africans	Brush-tailed porcupine	2		1	3
Erithizon dorsatum	American porcupine			3	3
Coendou prehensilis	Brazilian porcupine	1		2	3
Cavia porcellus	Domestic guinea pig	3	2	4	9
Kerodon rupestris	Rock cavy	2	3	5	10
Galea musteloides	Peruvian rock cavy			1	1
Dolichotis patagonium	Mara	1		2	3
Hydrochaerus isthmius	Capybara			2	2
Cuniculus paca	Paca			2	2
Dasyprocta leporina	Brazilian agouti	1		3	4
Myoprocta pratti	Acuchis			4	4
Lagostmus maximus	Plains viscachas			2	2
Myocastor coypus	Coypu	2	2	2	6
Capromys brownii	Brown's hutia			4	4
Capromys pilorides	Desmarest's hutia		1		1
Octodon degus	Degu			2	2
Proechimys guairae	Amazon spiny rat			1	1
Heterocephalus glaber	Naked mole rat	4	2	6	12
		27	18	99	144

ORDER LAGOMORPHA

Species	Common name	Baby	Junior	Adult	Total
Ochotona collaris	Collared pika			3	3
Oryctolagus cuniculus	European rabbit	2		4	6
Lepus europaeus	European brown hare			6	6
		2	0	13	15

ORDER TUBULIDENTATA

Species	Common name	Baby	Junior	Adult	Total
Orycteropus afer	Aardvark	5	0	1	6

APPENDIX 2   Comparative study by species of the cricothyroid area

Common name	Species	Weight (kg)	Area cricothyroid (cm²)
Tree hyrax	*Dendrohyrax arboreus*	1.33	0.8
Coati	*Nasua nasua*	2.8	1.0
Agouti	*Dasyprocta leporina*	3.42	2.2
Lar gibbon	*Hylobates lar*	6.0	1.9
Macaque	*Macaca irus*	6.0	1.7
Beaver	*Castor canadensis*	7.5	0.8
Oryx	*Oryx tao*	10.2	4.2
Muntjac	*Muntiacus muntjak*	10.36	1.0
Coyote	*Canis latrans*	13.2	2.9
Waterbuck	*Kobus defassa*	15.1	3.4
Wallaby	*Macropus parma*	16.0	Nil
Thomson's gazelle	*Gazelle thomsonii*	20.3	3.3
Mouflon	*Ovis orientalis*	21.1	3.71
Western roe deer	*Capreolus capreolus*	24.2	4.2
Cape hunting dog	*Lycaon pictus*	27.5	3.6
Lynx	*Lynx lynx*	28.3	3.08
Timor deer	*Cervus timorensis*	30.0	2.4
Collared peccary	*Tayassu tajacu*	30.1	4.5
Markhor	*Capra falconeri*	30.1	6.0
Puma	*Felis concolor*	32.4	2.9
Sika deer	*Cervus nippon*	40.0	4.2
Wolf	*Canis lupus*	40.4	6.8
Wart hog	*Phacochoerus aethiopicus*	45.3	7.9
Musk ox	*Ovibus moschatus*	50.4	11.1
Black bear	*Ursus americanus*	53.4	7.1
Sealion	*Zalophus californianus*	53.7	16.7
Cheetah	*Actinonyx jubatus*	55.2	13.2
Barbary sheep	*Ammotragus lervia*	55.2	8.4
Nyala	*Tragelaphus angasii*	60.5	4.97
Jaguar	*Panthera onca*	60.8	10.4
Reindeer/caribou	*Rangifer tarandus*	65.1	6.2
Brown bear	*Ursus arctos*	67.2	6.4
Giant panda	*Ailuropoda melanoleuca*	72.4	6.7
Wild boar	*Sus scrofa*	75.6	5.5
Fallow deer	*Cervus dama*	80.2	7.7
Chimpanzee	*Pan troglodytes*	80.2	6.6
Axis deer	*Cervus axis*	80.2	6.19
Fallow deer	*Cervus dama*	82.3	7.85
Jaguar	*Panthera onca*	85.1	13.7
Axis deer	*Cervus axis*	96.2	5.8

APPENDIX 2 (*cont.*)

Common name	Species	Weight (kg)	Area cricothyroid (cm^2)
Barbary sheep	*Ammotragus lervia*	100.1	5.9
Pere David's deer	*Elaphurus davidianus*	108.5	7.8
Guanaco	*Lama guanicoe*	110.0	5.9
Wilderbeast	*Connochaetus taurinus*	119.2	12.8
Gorilla	*Gorilla gorilla*	127.2	5.2
Zebra	*Equus burchelli*	130.3	8.3
Kudu	*Tragelaphus strepsiceros*	136.4	10.6
Tiger	*Panthera tigris*	142.7	9.0
Swamp deer	*Cervus duvaucelii*	148.5	8.3
Lion	*Panthera leo*	160.2	47.0
Congo buffalo	*Synceros caffer*	190.1	11.0
Sable antelope	*Hippotragus niger*	200.8	17.6
Gemsbok	*Oryx gazella*	205.0	5.9
Tapir	*Tapiris terrestris*	250.2	10.9
Roan antelope	*Hippotragus equinus*	250.3	13.2
Pigmy hippopotamus	*Choeropsis liberiensis*	252.4	21.8
Onager	*Equus hemionus*	260.3	7.2
Bison	*Bison bison*	265.0	29.1
Polar bear	*Thalarctos maritimus*	400.9	21.3
Bactrian camel	*Camelus bactrianus*	608.8	22.7
Yak	*Bos mutus*	800.0	14.1
Congo buffalo	*Synceros caffer*	815.0	27.0
Hippopotamus	*Hippotamus amphibius*	3800.0	14.1

APPENDIX 3   Comparative data from the glottic area and tracheal
ring broken down by age and sex

3.1   *Relationship between maximum glottic area and that within first tracheal ring*

Age	Sex	Name	Species	Glottic area (mm²)	Trachael area (mm²)
A	M	Alpaca	*Lama pacos*	228	239
A	F	Arabian camel	*Camelus dromedarius*	480	888
A	F	Bactrian camel	*Camelus bactrianus*	278	988
A	F	Bactrian camel	*Camelus bactrianus*	812	945
A	F	Bactrian camel	*Camelus bactrianus*	1149	1164
A	F	Bactrian camel	*Camelus bactrianus*	673	1046
A	F	Bactrian camel	*Camelus bactrianus*	240	416
B	F	Bactrian camel	*Camelus bactrianus*	72	83
A	F	Cape buffalo	*Synceros caffer*	327	1828
A	M	Cape buffalo	*Synceros caffer*	620	1423
A	F	Musk ox	*Ovibos moschatus*	1752	1286
A	M	Yak	*Bos mutus*	2089	2286
A	F	Yak	*Bos mutus*	1220	1272
A	F	Yak	*Bos mutus*	2071	2327
A	F	Blesbok	*Damaliscus dorcas*	105	247
A	F	Blesbok	*Damaliscus dorcas*	194	292
A	M	Blesbok	*Damaliscus dorcas*	199	413
A	M	Blesbok	*Damaliscus dorcas*	206	233
A	F	Greater kudu	*Tragelaphus strepsiceros*	247	397
A	F	Greater kudu	*Tragelaphus strepsiceros*	239	421
A	F	Greater kudu	*Tragelaphus strepsiceros*	286	586
A	M	Gemsbok	*Oryx gazella*	448	689
A	F	Gemsbok	*Oryx gazella*	1258	824
A	M	Nilgai	*Boselaphus tragocamelus*	387	388
A	F	Nilgai	*Boselaphus tragocamelus*	305	362
A	F	Nyala	*Tragelaphus angasii*	206	216
J	F	Roan antelope	*Hippotragus equinus*	133	199
A	F	Sable antelope	*Hippotragus niger*	385	606
A	F	American bison	*Bison bison*	923	265.2
A	M	American bison	*Bison bison*	685	270
A	M	American bison	*Bison bison*	256	142
A	F	Guanaco	*Lama guanico*	360	430
A	F	Blackbuck	*Antelope cervicapra*	136	268
A	F	Blackbuck	*Antelope cervicapra*	130	177
A	M	Blackbuck	*Antelope cervicapra*	196	203
B	M	Blackbuck	*Antelope cervicapra*	1.5	4.8
J	F	Waterbuck	*Kobus defassa*	159	659
J	F	Waterbuck	*Kobus defassa*	320	788

3.1 *Relationship between maximum glottic area and that within first tracheal ring (cont.)*

Age	Sex	Name	Species	Glottic area (mm^2)	Trachael area (mm^2)
J	M	Waterbuck	*Kobus defassa*	192	680
J	M	Waterbuck	*Kobus defassa*	339	433
A	M	Oryx	*Oryx tao*	260	653
A	F	Thomson's gazelle	*Gazelle thomsonii*	108	192
A	M	Fallow deer	*Cervus dama*	507	530
A	M	Fallow deer	*Cervus dama*	404	421
A	F	Timor deer	*Cervus timorensis*	324	337
A	M	Wart hog	*Phacochoerus aethipicus*	45	181
A	F	Collared peccary	*Tayassu tasucu*	78	181
A	M	Collared peccary	*Tayassu tasucu*	138	206
A	M	Collared peccary	*Tayassu tasucu*	115	250
A	F	Przewalski horse	*Equus przewalski*	394	687
J	M	Przewalski horse	*Equus przewalski*	625	1075
J	M	Przewalski horse	*Equus przewalski*	672	1119
A	M	Zebra	*Equus burchelli*	716	974
J	F	Zebra	*Equus burchelli*	702	962
A	F	Zebra	*Equus burchelli*	742	894
A	F	Onager	*Equus hemionus*	466	658
A	F	Cape hunting dog	*Lycaon pictus*	204	316
A	M	Wolf	*Canis lupus*	205	436
A	M	Wolf	*Canis lupus*	225	470
A	F	Wolf	*Canis lupus*	290	380
N	M	Wolf	*Canis lupus*	229	441
A	M	Cheetah	*Actinonyx subatus*	209	770
A	F	Cheetah	*Actinonyx subatus*	127	208
A	M	Cheetah	*Actinonyx subatus*	252	233
A	M	Cheetah	*Actinonyx subatus*	235	530
A	M	Cheetah	*Actinonyx subatus*	270	327
A	F	Cheetah	*Actinonyx subatus*	166	226
A	F	Cheetah	*Actinonyx subatus*	173	182
A	F	Lion	*Panthera leo*	776	1333
A	F	Lion	*Panthera leo*	902	1123
A	F	Lion	*Panthera leo*	505	632
A	F	Asiatic bear	*Selenarctos thibetanus*	110	862
A	F	Giant anteater	*Myrmecophaga tridactyla*	64	93
A	M	Red kangaroo	*Macropus rufus*	144	146
J	M	Red kangaroo	*Macropus rufus*	110	124
A	M	Wallaby	*Macropus rufogriseus*	43	98
A	F	Wallaby	*Macropus rufogriseus*	52	74
A	F	Wallaby	*Macropus rufogriseus*	65	73
A	F	Wallaby	*Macropus rufogriseus*	42	55
A	M	Wallaby	*Macropus rufogriseus*	24	34

3.2   *Relationship between maximum glottic and first tracheal ring diameters*

Age	Sex	Name	Species	Glottic diameter (mm)	Trachael diameter (mm)
A	M	Bactrian camel	*Camelus bactrianus*	21	38
A	M	Bactrian camel	*Camelus bactrianus*	13	23.50
A	F	Bactrian camel	*Camelus bactrianus*	15	29
A	M	Bactrian camel	*Camelus bactrianus*	14	35.5
A	M	American bison	*Bison bison*	25	48
A	F	Yak	*Bos mutus*	22	46
A	F	Greater kudu	*Tragelaphus strepsiceros*	18	20
A	F	Greater kudu	*Tragelaphus strepsiceros*	20	24
A	F	Guanaco	*Lama guanico*	22	22.5
A	M	Alpaca	*Lama pacos*	12	16
A	F	Nilgai	*Boselaphus tragocamelus*	17	21
A	F	Blesbok	*Damaliscus dorcas*	15	20
A	F	Blesbok	*Damaliscus dorcas*	12	19
A	F	Gemsbok	*Oryx gazella*	12	25
A	M	Cape buffalo	*Synceros caffer*	11	42
A	F	Cape buffalo	*Synceros caffer*	14	43
A	F	Collared peccary	*Tayassu tasucu*	11	16
A	F	Collared peccary	*Tayassu tasucu*	9	15
A	M	Waterbuck	*Kobus defassa*	16	25
A	F	Waterbuck	*Kobus defassa*	17	29
A	F	Waterbuck	*Kobus defassa*	11	26
A	F	Waterbuck	*Kobus defassa*	12	25.5
A	F	Blackbuck	*Antelope cervicapra*	11	26
A	F	Blackbuck	*Antelope cervicapra*	12.5	14
A	M	Blackbuck	*Antelope cervicapra*	10.5	12.5
A	M	Blackbuck	*Antelope cervicapra*	16.5	17
A	F	Blackbuck	*Antelope cervicapra*	10.4	10.5
A	F	Sable antelope	*Hippotragus niger*	16	34
J	F	Roan antelope	*Hippotragus equinus*	9	14
A	M	Musk ox	*Ovibus moschatus*	25.5	38
A	F	Przewalski horse	*Equus przewalski*	28	41
A	F	Przewalski horse	*Equus przewalski*	28	40
A	M	Przewalski horse	*Equus przewalski*	22	27.5
A	F	Przewalski horse	*Equus przewalski*	15.5	30.1
A	F	Onager	*Equus hemionus*	16	30
A	M	Zebra	*Equus burchelli*	22	26.5
A	F	Cheetah	*Actinonyx subatus*	14	18
A	F	Cheetah	*Actinonyx subatus*	16	12.5
A	F	Cheetah	*Actinonyx subatus*	10	17
A	F	Cheetah	*Actinonyx subatus*	14	17
A	F	Cheetah	*Actinonyx subatus*	14	16

3.2 *Relationship between maximum glottic and first tracheal ring diameters*

Age	Sex	Name	Species	Glottic diameter (mm)	Trachael diameter (mm)
A	M	Cheetah	*Actinonyx subatus*	16	16
A	M	Wolf	*Canis lupus*	18	20
A	F	Wolf	*Canis lupus*	13	20
J	F	Wolf	*Canis lupus*	14	20
A	F	Wolf	*Canis lupus*	11	19.75
A	F	Cape hunting dog	*Lycaon pictus*	12.	18.5
A	M	Cape hunting dog	*Lycaon pictus*	20	23
A	F	Lion	*Panthera leo*	18	26
A	F	Lion	*Panthera leo*	19	34.5
A	F	Lion	*Panthera leo*	14.5	30.2
J	M	Wallaby	*Macropus rufogriseus*	5	5
J	M	Wallaby	*Macropus rufogriseus*	10	10
A	F	Wallaby	*Macropus rufogriseus*	8	8
A	F	Wallaby	*Macropus rufogriseus*	7	10
A	F	Wallaby	*Macropus rufogriseus*	7	8
A	M	Wallaby	*Macropus rufogriseus*	10	10
A	M	Red kangaroo	*Macropus rufus*	8	10.3

APPENDIX 4   Measurements of cricoarytenoid joint facets

Common name	Dimensions (mm) arytenoid facet		Dimensions (mm) cricoid facet	
	Right	Left	Right	Left
Yak	12.5 × 11.5	12.5 × 11.5	14.5 × 9.0	14.5 × 9.0
Wild boar	7.5 × 6.0	7.5 × 6.0	7.0 × 4.5	7.0 × 4.5
Cheetah	6.0 × 3.5	6.0 × 4.0	7.0 × 4.0	6.0 × 3.5
Cheetah	5.0 × 4.0	5.0 × 4.0	6.0 × 4.0	6.0 × 4.0
Congo buffalo	1.3 × 1.0	1.2 × 1.0	1.6 × 0.9	1.6 × 0.85
Cape buffalo	1.6 × 1.2	1.6 × 1.2	1.8 × 0.9	1.8 × 0.9
Przewalski horse	13.0 × 8.5	12.5 × 9.0	13.0 × 6.5	13.5 × 6.5
Przewalski horse	12.0 × 8.5	11.5 × 9.0	12.0 × 7.0	12.0 × 6.5
Cape hunting dog	3.6 × 2.4	3.8 × 2.2	4.8 × 1.6	4.4 × 1.6
Wolf	7.5 × 7.0	8.0 × 6.5	12.5 × 4.5	12.0 × 5.0
Wolf	8.0 × 6.0	7.5 × 6.0	8.5 × 4.0	8.5 × 4.5
North American bison	1.1 × 0.8	1.0 × 0.8	1.2 × 0.7	1.2 × 0.7
North American bison	1.05 × 0.8	0.95 × 0.95	1.2 × 0.8	1.2 × 0.7
Zebra	11.0 × 8.0	10.5 × 10.0	13.0 × 6.0	12.5 × 5.0
Onager	13.5 × 11.5	14.5 × 11.5	13.5 × 10.0	14.0 × 10.5
Onager	12.5 × 11.5	12.0 × 11.0	13.0 × 8.5	15.5 × 9.5
Kudu	8.0 × 6.5	8.0 × 5.5	9.5 × 5.0	9.5 × 4.5
Kudu	7.0 × 5.5	7.0 × 5.5	9.5 × 5.5	10.5 × 6.0
Nilgai	9.5 × 9.0	10.0 × 9.5	8.5 × 4.5	8.5 × 4.5
Musk ox	1.2 × 1.1	1.2 × 1.3	1.8 × 0.8	1.5 × 0.8
Lion	10.0 × 5.0	10.0 × 5.0	11.0 × 6.5	11.4 × 6.0
Collared peccary	2.0 × 1.5	2.0 × 1.5	3.5 × 2.5	3.5 × 2.0

Index